Elementary Agriculture

ABOUT THE AUTHORS

Dr. Kanchan Nainwal is Associate Professor/Associate Director, Agronomy at Krishi Vigyan Kendra, Almora, G B Pant University of Agriculture & Technology, Pantnagar. Prior to this she was at Agriculture Technology Information Centre (ATIC), Pantnagar involved in extension and teaching students for nine years. She has worked in the Directorate of Sugarcane Development, Ministry of Agriculture and Farmer's Welfare from the Year 2001-2003. She obtained her graduation in Agriculture & Animal Husbandry, Master's Degree and PhD Degree from Agronomy Department, College of Agriculture with all first class from G B Pant University of Agriculture & Technology, Pantnagar. Dr. Nainwal has experience in Teaching, Research and Extension work in Agronomy in various capacities for 17 years. She has written 235 papers including research papers, presented in conference and symposiums, online publications, popular articles, extension booklets and 2 books and 14 book chapters.

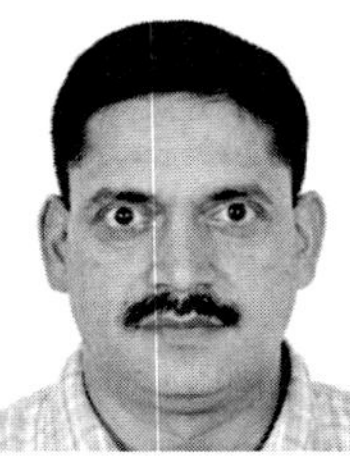

Dr. Navin Chandra Nainwal is presently Consultant, RKVY in Ministry of Agriculture and Farmer's Welfare, Government of India. He is Alumni of G.B. Pant University of Agriculture and Technolgy, Pantnagar and his specialization (M.Sc. Agril. & PhD.) is in post harvest technology and supply chain management. Dr. Nainwal has done Advanced Management Programme - Developing Corporate leadership in Challenging Environment" at Administrative Staff College of India (ASCI) Hyderabad and in **Europe-Italy, France and Switzerland**. He got the Netherland Government Nuffic Scholarship for Practical Management Course on Vegetable Production at PTC + International, Ede, the **Netherlands** and prestigious MASHAV fellowship for International R & D Course in Protected Cultivation at Volcani Research Centre, ARO, **Israel.** He was Assistant Director at Amity University Uttar Pradesh and established Amity Institute of Horticulture Studies & Research, in the Year 2010. Prior to this he was Coordinator In-charge for Agri Export Development Unit, Dehradun Government of Uttarakhand. He has 25 years long experience, worked with World Bank Project, Food & Agriculture Organizations of United Nations (FAO) and Bill and Melinda Gates Foundation-WFLO Project. He has written 40 papers and 7 books and many popular articles. He has presented research papers in abroad at various international forums. He is Hon' General Secretary of Walnut and Other Nut Fruit Crop Growers Association of India (WANGAI), Member, Regional Committee on Agriculture & Food, CII Northern Region and Founder Member, Asian Food Safety & Security Association (AFSSA), Bangladesh.

Dr. A. Arunachalam is a trained ecologist from the North-Eastern Hill University, Shillong and has grown professionally through strong pursuit in ecological research that is evident from 11 books, 180 research papers, 110 book chapters, 25 popular articles. He started his service as a Lecturer/Assistant Professor in Forestry, North Eastern Regional Instituteof Science & Technology (Arunachal Pradesh) and currently he is working as a Principal Scientist in the headquarters of Indian Council of Agricultural Research (ICAR), KrishiBhawan, New Delhi. His academic excellence is also evident from the awards/honours he has received: DST-Visting Fellowship in Biodiversity, Jawaharlal Nehru Centre for Advanced Study in Science & Technology, Jakkur, 1998-99; DST-BOYSCAST Fellowship in Restoration Ecology, University of Massachusetts, Boston (USA), 2000-01; Dr. K.G. Tejwani Award for Management of Natural Resources (biennium 2002-03), Indian Association of Soil and Water Conservationists; Pran Vohra Award in Agriculture and Forestry Sciences (2003-04), Indian Science Congress Association; Eminent Scientist of the Year award (2012) by the National Environmental Science Academy and several others; Dr. Arunachalam is a Fellow of several professional societies including National Academy of Agricultural Sciences, National Academy of Dairy Sciences (India) and National Academy of Biological Sciences. Presently, he is the Vice President of the Society for Science in Climate Change and Sustainable Environment, New Delhi and is the Chief Editor of Indian Journal of Hill Farming.

Elementary Agriculture

Kanchan Nainwal
Associate Director, Agronomy
(KVK-Almora)
G.B. Pant University of Agriculture & Technology
Pantnagar (Uttarakhand)

Navin Nainwal
Consultant, RKVY
Ministry of Agriculture and Farmers Welfare
Kirishi Bhawan, New Delhi - 110001

A Arunachalam
Principal Scientist
Indian Council of Agricultural Research
New Delhi - 110001

NEW INDIA PUBLISHING AGENCY
New Delhi – 110 034

NEW INDIA PUBLISHING AGENCY
101, Vikas Surya Plaza, CU Block, LSC Market
Pitam Pura, New Delhi 110 034, India
Phone: + 91 (11)27 34 17 17 Fax: + 91(11) 27 34 16 16
Email: info@nipabooks.com
Web: www.nipabooks.com

Feedback at feedbacks@nipabooks.com

ISBN: 978-93-85516-53-5

Composed, Designed and Printed in India

रानी लक्ष्मी बाई केन्द्रीय कृषि विश्वविद्यालय
ग्वालियर रोड, झांसी 284 003 (उत्तर प्रदेश) भारत
Rani Lakshmi Bai Central Agricultural University
Gwalior Road, Jhansi 284 003 (U.P.) India

Phone : 0510-2730777
Telefax : 0510-2730555
E-mail : vcrlbcau@gmail.com
Website : www.rlbcau.ac.in

डॉ अरविन्द कुमार
कुलपति
Dr Arvind Kumar
Vice-Chancellor

Foreword

Indian agriculture has moved from the era of shortage to surplus owing largely to high-yielding crop varieties and improved crop management technologies. Today, the country acknowledges the role of agricultural scientists for their contribution to the increase in food production. Nonetheless, the challenge is to sustain agricultural production and productivity for the growing human population and to make our system climate resilient. Herein, the role of youth is important, as they need to consider agriculture as a profitable profession. In this context, the book titled 'Elementary Agriculture' has comprehensive information on fundamental topics in agriculture that will be highly beneficial to the undergraduate students.

I commend the authors for having adopted simple language to augment available technical information in Indian agriculture for better understanding of the students.

(ARVIND KUMAR)

Dated the 1st December, 2016
Jhansi, Uttar Pradesh

Camp Office: Room No. 213, KAB-II, Pusa, New Delhi-110012 *Ph:* (O) 011-25846034 *Fax:* 011-25843932

Preface

Agriculture is the backbone of our economy, where 58 % population is dependent on agriculture for their livelihood. The whole philosophy of agriculture revolves around the principle of successful management of various inputs for crop production to satisfy changing human needs while maintaining or enhancing the quality of environment and conserving natural resources. Therefore, authors had tried their best in including each aspects of agriculture.

This book has been mainly designed especially for the undergraduate students in agriculture, agricultural engineering and other allied subjects. This book is entirely based on the syllabus of agriculture in different universities and is written in very simple and understandable words. This book includes almost all the topics i.e. History of agriculture, soils, fertilizers and manures, irrigation and drainage and rainfed agriculture. Based on different examinations held in the university, one chapter on question-answer is also developed, which is also very informative for the students appearing in different competitions.

The authors have attempted to collect information from various sources to compose this text book in the larger interest of students' readership. With additional information, as would be available in the time to come, this edition would be appropriately revised and updated

Authors

Contents

Chapter 1

Indian Agriculture: Scope and Resources

Agriculture is a branch of Applied science. It is the main enterprise in the world. In a true sense, it is a productive unit where the gift of nature i.e. land, light, air, temperature, water, humidity etc. are integrated into a single primary unit – crop plants and their usable parts are thus indispensable for human beings. Secondary productive units are animals (livestock), birds, insects etc. and they feed on these primary units and provide more solid products such as meat, milk, wool, eggs, honey, silk, leather etc.

The word agriculture is derived from two Latin words i.e. Ager/Agric means soil and cultura means cultivation. Thus, *"Agriculture is a science of cultivation of land"*. It is an art and science both in which we study all the activities of human beings related to use of soil.

Or

Agriculture in its widest sense can be defined as the cultivation and/or production of crops plants or livestock products.

Or

Agriculture is an activity of man primarily aimed at the production of food, fibre, fuel etc. by optimum use of terrestrial resources i.e. land and water (ICAR, 1996).

Presently, the definition of Agriculture is extended to scientific production techniques combined with developing improved varieties of crops, trees, crop protection, soil reclamation, animal sciences, milk production, fisheries, poultry production and evolution of manures and fertilizers.

Agriculture can be termed as a science, an art and business altogether.

Science: Because it provides new and improved strain of crop and animal with the help of the knowledge of breeding and genetics, modern technology of dairy science.

Art: Because it is the management whether it is crop or animal husbandry.

Commerce (Business): Because the entire agricultural produce is linked with marketing, which brings in the question of profit or loss.

Scope of Agriculture

About 70% of India's population live in villages. The occupation of villagers is agriculture. Agriculture is the dominant sector of our economy and contributes in various ways such as:

- **Food Supply:** The food grain production of India is 252.68 million tons in 2014-15 (Economic Survey, 2014-15) to feed the growing population of India. India, thus, is able to meet almost all the needs of its population with regards to food by developing intensive program for increasing food production.
- **National Economy:** Agriculture sector accounts for 15% of the national GDP, about 11% of its exports and half of the population still rely on agriculture.
- **Total Employment:** Nearly 58% of the rural population earns its livelihood from agriculture and other occupation allied to agriculture. In cities also, a considerable part of labour force is engaged in jobs depending on processing and marketing of agricultural products.
- **Industrial Inputs:** Most of the industries depend on the raw material produced by agriculture, so agriculture is the principal source of raw material to the industries. The industries like cotton textile, jute, paper, sugar depends totally on agriculture for the supply of raw material. The small scale and cottage industries like handloom , ginning and pressing, oil crushing, rice husking, sericulture, fruit processing, etc are also mainly agro based industries.
- **State Revenue:** The agriculture is contributing the revenue by agriculture taxation includes direct tax and indirect tax. Direct tax includes land revenue, cesses and surcharge on land revenue, cesses on crops and agricultural income tax. Indirect tax induces sales tax, custom duty and local octri, etc. which farmer pay on purchase of agriculture inputs.
- **Trade:** Agriculture plays an important role in foreign trade attracting valuable foreign exchange, necessary for our economic development. The

product from agriculture based industries such as jute, cloth, tinned food, etc. contributed to 20% of our export. Around 50% of total exports are contributed by agricultural sector. Indian agriculture plays an important role in roads, rails and waterways outside the countries. In India roads, rails and waterways used to transport considerable amount of agricultural produce and agro based industrial products. Agricultural products like tea, coffee, sugar, oil seeds, tobacco; spices, etc. also constitute the main items of export from India.

History of Agriculture

Inspite of numerous publication, we may not exactly know the origin of cultivated plants, their diffusion paths and ancient agricultural systems. It is supposed that man has his existence on the earth about 15 lakh years ago. The man was evolved from monkey who started to move by standing erect on feet, therefore called Homo erectus or Java man. Later on Java man was transferred into cro-magnan into modern man, as we are today. Such man *(Homo sapiens),* developed learning habit, and evolved first in Africa about 35000 years ago. In the beginning, man started to pet dog (the first pet animal), sheep and goat.

Western India

Western Asia (Israel, Jordan, Iraq etc.) is considered to be the birth place of agricultural revolution, where wild ancestors of wheat and barley and domestic animals like goat, sheep, pig and cattle are found. The period from 7500-6500 B.C. was in real sense the discovery of agriculture.

The period from 3000-1700 B.C. market the spread of agricultural revolution to Egypt and subsequently to Indian valley, which took place in countries between river Nile and Ganges. Man invented plough (for using with the help of oxen) and stone implements were supplemented with Copper and Bronze. The centre of agricultural revolution in Indus Valley was from Mohenjodaro to Harrappan. Then, Indus Valley civilisation spread to Punjab, Haryana, Jammu, Uttar Pradesh, Rajasthan, Gujarat and Madhya Pradesh. Harrappans raised bread wheat, barley, sesame, peas, melons, cotton, rape and mustard. Animals like cattle, buffalo, goat, sheep, pig, camel, ass, dog and cat were domesticated. Harrappan agriculture also spread to Andhra Pradesh, Karnataka and adjoining areas.

About 1800-1600 B.C., Aryans migrated to India and over whelmed the Harrappans. Horses was main domesticated animals besides cattle. Agriculture was very important profession during *Vedic age* (1500-1000 B.C.). Iron implements became common. References were present in Vedas during this period besides barley, wheat, beans, sesame, millets and rice.

Buddhist period (600 B.C.) marks the importance of trees. It can be called as a period of Arboriculture (forestry) and Horticulture. Only wheat and barley (Winter) and rice and millets (Summer) were common.

South India

During the first century of Christian Era and 300 A.D., the most important development in agriculture was cultivation of rice in South India, where Cauvery river was main source of irrigation. Crops like Rice, finger millet, sugarcane, pepper and turmeric were taken. After 300 AD Gupta period contributed to agriculture when *Amarkosha* written, contains all the information on soils, irrigation, implements and land use.

During Kannauj Empire of Harsha (606-647 A.D.) a rice variety of large grains of extraordinary fragrance called 'rice of grandees' was grown in Magadh.

Krishi Parashara which was written by Parashara (950-1100 A.D.) contains history of agriculture from ninth to eleventh century. It includes land use, manuring, crop rotation, irrigation, tillage, implements, crop protection, and agricultural meteorology. Anantaraja Sagar (Porumamilla tank) in Cuddapah district of Andhra Pradesh was build in 1367 by Vijaynagar King. He constructed the great dam and channel of Korragal and the Basavanna channel. Rice, wheat, sorghum, barley, beans, greengram, horsegram etc. were grown. During 1336-1646 A.D., Bahmanis constructed several canals such as Utpalpur Nandashaila, Bifbihra etc. During 1498-1580, Portuguese introduced new world plants to India. Groundnut, tobacco, potato, cashew, guava, pineapple, chillies were introduced.

Agriculture during British Rule and in Free India

Zamindari System: It is system in which land ownership was given to small groups of people to collect rent from individual farmers and pay to government.

Ryotwari System: It is the system in which the rulers collect rent directly from farmers who had been settled on the land.

Agricultural Development

On 27th April, 1871, a joint department of agricultural revenue and commerce was established by Lord Mayo on the request of A.O. Hume. As per direction of famine commission 1880, a seperate department of Agriculture was set up in 1881.

IARI: In 1905, Imperial Agricultural Research Institute was established at Pusa (at that time Bengal but presently in Bihar) under the Viceroyalty of Lord

Curzon. But due to earthquake in 1934, it was shifted to Pusa Road, New Delhi. After independence, the word 'Imperial' was substituted by 'Indian' and now it is called *Indian Agricultural Research Institute*. In 1958, IARI was given the status of deemed university by University Grant Commission (UGC).

ICAR: In 1926, under the Viceroyalty of Lord Linlithgow, the Royal Commission on Agriculture was constituted. In 16th July 1929, *Imperial Council of Agricultural Research* was established. First President of ICAR was Mohammad Habibbullah. Later in March 1946, under the Presidentship of Joginder Singh, the ICAR was renamed as *Indian Council of Agricultural Research*. It is the central body, established for improvement and coordination of Agricultural research throughout the country. In 1966, ICAR became full autonomous body and its first Director General was Dr. B.P. Pal.

History of Indian Agriculture - Historic Developments

The historic developments in agriculture during British rules and free India are:

Sr. No.	Year	Historic Developments
1.	1871	Departments of Agriculture Created.
2.	1878	Higher Education in Agriculture at Coimbatore.
3.	1880	Famine Commission Appointed.
4.	1890	Higher Education in Agriculture at Pune.
5.	1891	Dr. J A Voekker Report on Improving Indian Agriculture.
6.	1900	Forest Research Institute.
7.	1901	First Immigration Commission.
8.	1905	Imperial (now Indian) Agricultural Research Institute at Pusa (Now at Delhi).
9.	1921	Indian Central Cotton Committee.
10.	1926	Royal Commission on Agriculture Headed by Lord Linlithgow.
11.	1929	Imperial (now Indian) Council of Agricultural Research at Delhi.
12.	1936	Indian Central Jute Committee.
13.	1942	Department of Food Created.
14.	1942	Grow More Food Campaign.
15.	1944	Indian Central Sugarcane Committee.
16.	1945	Indian Central Tobacco Committee.
17.	1946	Directorate of Plant Projection & Quarantine.
18.	1946	Central Rice Research Institute.
19.	1947	Food Policy Committee.
20.	1947	Fertilizers & Chemicals Travancore.
21.	1956	Project for Intensification of Regional Research on Cotton, Oil Seeds, Millets (PIRRCOM).
22.	1957	All India Coordinated Maize Improvement Project.
23.	1960	Intensive Agriculture District Programme (IADP).
24.	1960	First Agricultural University at Pant Nagar.
25.	1963	National Seed Corporation.
26.	1965	Intensive Agriculture Area Programme (IAAP).

contd...

Sr. No.	Year	Historic Developments
27.	1965	National Demonstration Programme.
28.	1966	High Yielding Varieties Programme.
29.	1966	Directorate of Extension.
30.	1966	Multiple Cropping Schemes.
31.	1969	Second Immigration Compassion.
32.	1970	Drought Prone area Programme (DPAP).
33.	1970	National Commission on Agriculture.
34.	1971	All India Coordinated Project for Dry Land Agriculture.
35.	1972	ICRISAT.
36.	1973	Minikit Trials Programme.
37.	1974	Command Area Development.
38.	1976	Integrated Rural Development Programme (IRDP).
39.	1977	Training & Visit System (T&V).
40.	1979	National Agricultural Research Project (NARP).
41.	1982	National Bank for Agriculture & Rural Development (NABARD).
42.	1985	National Agricultural Extension Project (NAEP).
43.	1986	National Agricultural Research Project (Phase-II).
44.	1990	National Agricultural Technology Project (NATP).

Agricultural Colleges and Universities

During 1901-1905, six agricultural colleges were established in India at Poona, Kanpur, Sabour, Nagpur, Loyalpur (now in Pakistan) and Coimbatore. The course study of Agricultural Extension was first started at Sabour (Bhagalpur, Bihar).

On the recommendation of Joint Indo-American team, first Agriculture University was established at Pantnagar in 1960 (Chief Minister of Uttar Pradesh was Pt. Govind Ballabh Pant at that time). At present, in India, there are 45 State Agriculture Universities and 4 Central Universities.

List of Agricultural Universities in India

S.No. Name of University

State Agricultural Universities/Central Universities

1. Acharya NG Ranga Agricultural University, Rajendranagar, Hyderabad-500030, A.P.
2. Anand Agricultural University, Anand-388110, Gujarat
3. Assam Agricultural University, Jorhat-785013, Assam
4. Bidhan Chandra Krishi Viswavidyalaya, P.O Krishi Viswavidyalaya, Mohanpur, Nadia-741252, West Bengal

5. Birsa Agricultural University, Kanke, Ranchi- 834006, Jharkhand
6. Central Agricultural University, Imphal-795004, Manipur
7. Chandra Shekar Azad University of Agriculture & Technology, Kanpur-208002 Uttar Pradesh.
8. Chaudhary Charan Singh Haryana Agricultural University, Hissar-125004 Haryana
9. Chaudhary Sarwan Kumar Himachal Pradesh Krishi Vishwavidyalaya Palampur, Kangra-176062, Himachal Pradesh
10. Dr Balasaheb Sawant Konkan Krishi Vidyapeeth, Dapoli, Ratnagiri-415712 Maharashtra
11. Dr Panjabrao Deshmukh Krishi Vidyapeeth, Krishi Nagar, Akola-444104 Maharashtra
12. Dr Yashwant Singh Parmar University of Horticulture & Forestry, Solan Nauni-173230, Himachal Pradesh
13. Govind Ballabh Pant University of Agriculture & Technology, Pantnagar Udhamsingh Nagar-263145, Uttarakhand
14. Guru Angad Dev University of Veterinary and Animal Sciences Ludhiana-141004, Punjab
15. Indira Gandhi Krishi Vishwavidyalaya, Krishak Nagar, Raipur-492006 Chhattisgarh
16. Jawaharlal Nehru Krishi Vishwavidyalaya, Krishi Nagar, Jabalpur- 482004 Madhya Pradesh
17. Junagadh Agriculture University, Moti Baug, Agril. Campus, Junagadh-362001, Gujarat
18. Karnataka Veterinary Animal and Fisheries Science University, P.B. No.6 Nandinagar, Bidar-585401, Karnataka
19. Kerala Agricultural University, P.O Vellanikkara, Thrissur-680656, Kerala
20. Maharana Pratap University of Agriculture & Technology, Udaipur-313001 Rajasthan
21. Maharashtra Animal Science & Fishery University, Nagpur, Maharashtra
22. Mahatma Phule Krishi Vidyapeeth, Rahuri-413722, Maharashtra
23. Marathwada Agricultural University, Parbhani-431402, Maharashtra

24. Narendra Deva University of Agriculture & Technology, Kumarganj, Faizabad-224229, Uttar Pradesh
25. Navsari Agricultural University, Vijalpore, Navsari-396450, Gujarat
26. Orissa University of Agriculture & Technology, Siripur, Bhubaneswar-751003 Odisha
27. Punjab Agricultural University, Ludhiana-141004, Punjab
28. Rajasthan Agricultural University, Bikaner-334006, Rajasthan
29. Rajendra Agricultural University, Pusa, Samastipur-848125, Bihar
30. Sardar Vallabh Bhai Patel Univ of Agriculture & Technology, Modipuram, Meerut-250110, Uttar Pradesh
31. Sardarkrushinagar-Dantiwada Agricultural University, Sardarkrushinagar, Dantiwada, Banaskantha-385506, Gujarat
32. Sher-e-Kashmir University of Agricultural Sciences & Technology, Railway Road, Jammu-180012 (J&K)
33. Sher-E-Kashmir University of Agricultural Sciences & Technology, Shalimar, Srinagar-191121, (J&K)
34. Sri Venkateswara Veterinary University, Tirupati, Chittoor-517502, Andhra Pradesh
35. Tamil Nadu Agricultural University, Coimbatore-641003, Tamil Nadu
36. Tamil Nadu Veterinary & Animal Sciences University, Madhavaram Milk Colony, Chennai- 600051, Tamil Nadu
37. University of Agricultural Sciences, Dharwad, Karnataka
38. University of Agricultural Sciences, Bengaluru-560065, Karnataka
39. UP Pandit Deen Dayal Upadhaya Pashu Chikitsa Vigyan Vishwavidyalaya Evam, Go Anusandhan Sansthan, Mathura-281001, Uttar Pradesh
40. Uttar Banga Krishi Vishwavidyalaya, P.O. Pundibari, Distt. Cooch Behar-736165, West Bengal
41. West Bengal University of Animal & Fishery Sciences, 68 KB Sarani, Kolkata-700037, West Bengal
42. University of Horticultural Sciences, Venkataramnagudem, West Godavari, Andhra Pradesh
43. Rajmata VRS Agricultural University, Gwalior-474002, Madhya Pradesh

44. University of Horticultural Sciences, Navanagar, Bagalkot-587101 Karnataka
45. University of Agricultural Sciences, Raichur-584102, Karnataka

Deemed-to-be Universities

1. Indian Agricultural Research Institute, Pusa-110012, New Delhi
2. Indian Veterinary Research Institute, Izatnagar, Bareilly-243122, Uttar Pradesh
3. National Dairy Research Institute, Karnal-132001, Haryana
4. Central Institute of Fisheries Education, Mumbai-400061, Maharashtra
5. Allahabad Agricultural Institute, Allahabad-211007, Uttar Pradesh

Central Universities with Agriculture Faculty

1. Banaras Hindu University, Varanasi, Uttar Pradesh
2. Aligarh Muslim University, Aligarh, Uttar Pradesh
3. Vishwa Bharti, Shantiniketan, West Bengal
4. Nagaland University, Mediziphema, Nagaland

ICAR Institutes

1. Indian Agricultural Research Institute, New Delhi
2. Indian Agricultural Statistics Research Institute, New Delhi
3. National Academy of Agricultural Research Management, Hyderabad, Andhra Pradesh
4. Central Research Institute for Dryland Agriculture, Hyderabad, Andhra Pradesh
5. Central Soil Salinity Research Institute, Karnal, Haryana
6. National Dairy Research Institute, Karnal, Haryana
7. Central Avian Research Institute, Bareilly, Uttar Pradesh
8. Indian Veterinary Research Institute, Bareilly, Uttar Pradesh
9. Central Institute for Cotton Research, Nagpur, Maharashtra
10. Central Institute for Research on Buffalo (CIRB), Hisar

11. Central Institute for Research on Cotton Technology, Mumbai, Maharashtra
12. Central Institute of Fisheries Education, Mumbai
13. Central Institute for Research on Goats, Mathura, U.P.
14. Indian Institute of Sugarcane Research, Lucknow, U.P.
15. Central Institute for Subtropical Horticulture, Lucknow, U.P.
16. Central Institute of Agricultural Engineering, Bhopal, M.P.
17. Indian Institute of Soil Science, Bhopal, Madhaya Pradesh
18. Central Institute of Arid Horticulture, Bikaner, Rajasthan
19. Central Institute of Brackishwater Aquaculture, Chennai
20. Central Institute of Post Harvest Engineering & Technology, Ludhiana, Punjab
21. Central Institute of Temperate Horticulture, Srinagar, J&K
22. Central Marine Fisheries Research Institute, Cochin, Kerala
23. Central Institute of Fisheries Technology, Cochin, Kerala
24. Central Plantation Crops Research Institute, Kasaragod, Kerala
25. Central Arid Zone Research Institute, Jodhpur, Rajasthan
26. Central Potato Research Institute, Shimla, H.P.
27. Central Research Institute for Jute and Allied Fibres, Barrackpore, West Bengal
28. Central Inland Fisheries Research Institute, Barrackpore, West Bengal
29. Central Rice Research Institute, Cuttack, Odisha
30. Central Sheep and Wool Research Institute, Avikanagar, Rajasthan
31. Central Soil and Water Conservation Research and Training Institute Dehradun, Uttarakhand
32. Central Tobacco Research Institute, Rajahmundry, Andhra Pradesh
33. Central Tuber Crops Research Institute, Thiruvananthapuram, Kerala
34. ICAR Research Complex for Eastern Region, Patna, Bihar
35. ICAR Research Complex for Goa, Goa

36. ICAR Research Complex for NEH Region, Ri Bhoi, Meghalaya
37. Central Agricultural Research Institute, Port Blair
38. Indian Grassland & Fodder Research Institute, Jhansi, U.P.
39. Indian Institute of Horticultural Research, Bengaluru, Karnataka
40. National Institute of Animal Nutrition and Physiology, Bengaluru, Karnataka
41. Indian Institute of Natural Resins and Gums, Ranchi
42. Indian Institute of Pulses Research, Kanpur, U.P.
43. Indian Institute of Spices Research, Marikunnu, Kerala
44. Indian Institute of Vegetable Research, Varanasi, U.P.
45. National Institute of Research on Jute and Allied Fibre Technology, Kolkata, West Bengal
46. Sugarcane Breeding Institute, Coimbatore, Tamil Nadu
47. Vivekananda Parvatiya Krishi Anusandhan Sansthan, Almora, Uttarakhand
48. Central Institute of Freshwater Aquaculture, Bhubaneshwar, Odisha

Bureau

1. Agriculturally Important Micro Organisms, Mau Nath Bhanjan, Uttar Pradesh
2. Animal Genetic Resources, Karnal, Haryana
3. Fish Genetic Resources, Lucknow, Uttar Pradesh
4. Plant Genetic Resources, New Delhi
5. Soil Survey & Land Use Planning, Nagpur, Maharashtra

Project Directorate

1. Maize, New Delhi
2. Seed, Mau Nath Bhanjan, Uttar Pradesh
3. Wheat, Karnal, Haryana
4. Biological Control, Bengaluru, Karnataka
5. Animal Disease Monitoring and Surveillance, Bengaluru, Karnataka
6. Cropping Systems, Meerut, Uttar Pradesh

7. Cattle, Meerut, Uttar Pradesh
8. Foot and Mouth Disease, Mukteshwar, Uttarakhand
9. Poultry, Hyderabad, Andhra Pradesh
10. Oilseeds, Hyderabad, Andhra Pradesh
11. Rice, Hyderabad, Andhra Pradesh

National Research Centre

1. Agricultural Economics and Policy, New Delhi
2. Integrated Pest Management, New Delhi
3. Plant Biotechnology, New Delhi
4. Agroforestry, Jhansi, Uttar Pradesh
5. Banana, Thiruchirapalli, Tamil Nadu
6. Cashew, Dakshina Kannada, Karnataka
7. Soybean, Indore, Madhya Pradesh
8. Citrus, Nagpur, Maharashtra
9. Equines, Hisar, Haryana
10. Grapes, Pune, Maharashtra
11. Onion and Garlic, Pune, Maharashtra
12. Groundnut, Junagarh, Gujarat
13. Litchi, Muzaffarpur, Bihar
14. Medicinal and Aromatic Plants, Anand, Gujarat
15. Mushroom, Solan, Himachal Pradesh
16. Oilpalm, West Godavari, Andhra Pradesh
17. Orchids, Gangtok, Sikkim
18. Sorghum, Hyderabad, Andhra Pradesh
19. Meat, Hyderabad, Andhra Pradesh
20. Weed Science, Jabalpur, Madhya Pradesh
21. Women in Agriculture, Bhubaneshwar, Odisha
22. Water Technology Center for Eastern Region, Bhubaneshwar, Odisha

23. Camel, Bikaner, Rajasthan
24. Mithun, Dimapur, Nagaland
25. Pig, Guwahati, Assam
26. Pomegranate, Solapur, Maharashtra
27. Rapeseed-Mustard, Bharatpur, Rajasthan
28. Seed Spices, Ajmer, Rajasthan
29. Cold Water Fisheries, Bhimtal, Uttarakhand
30. Yak, West Kameng, Arunachal Pradesh

References

Randhawa, N.S. 1983. History of Agriculture in India Vol III. 1757-1947 ICAR, New Delhi.
Reddy, S.R. Principles of Agronomy 1999. Kalyani Publisher.

Chapter 2

Cropping Seasons and Classification of Crops

What is Crop?

Crop refers to plants sown and harvested by man for economic purposes.

Cropping Seasons: There are three main seasons and two sub seasons.

1. ***Rainy or Kharif Season*** – Main characteristics of the Kharif season are
   - It starts from June and spreads to September.
   - There is high temperature and high humidity.
   - More than 2/3rd of annual rainfall is received in a year.
   - In the beginning, days are longer than nights while days and nights are equal when season ends.
   - Important crops grown are maize, rice, sorghum, Moong, Urd, groundnut, cowpea, soybean etc.
2. ***Winter or Rabi Season*** – Common characteristics are
   - It starts from October – November and ends in March.
   - There is decline in temperature from November to January but from February onwards, from there is rise in temperature.
   - There is scanty rainfall. In winter, rains are about 2% of the total rain.
   - In the beginning, days and nights are equal but as winter becomes severe, days turns shorter and nights longer. However, at the end of season again, days and nights are equal.
   - Important crops grown are wheat, oat, barley, barseem, pea, lentil etc.

3. ***Summer or Zaid Season*** – Main characteristics are

- It starts from April and ends in June.
- There is high temperature, low humidity and hot dry winds.
- No rainfall occurs in this season.
- Days are long and nights are short.
- Crops grown are urd, moong, cowpea, maize, cucumber, muskmelon, watermelon etc.

Sub seasons

a. ***Autumn season***: Transit between rainy and winter season.

- It is short season, and confines from October to November.
- Here, rainy season departs and winter season arrives.
- There is neither warm nor cold during night and days are pleasant.
- Crops grown are Sugarcane, Potato, Lahi etc.

b. ***Spring season:*** Transit between winter and summer season.

- Days starts enlarging and same with temperature and nights starts shrinking.
- Temperature is very pleasant.
- Crops grown are Sugarcane, Maize, Cucurbits, Summer pulses etc.

Classification of Crops

Different Basis of Classification of Crops: There are different basis of classifying crops:

A. According to Life cycle of Plants

i. ***Annuals***: Annuals are those crops which complete their life cycle in one or less than one year. e.g. Wheat, rice, maize, barley, bajra, urd, moong etc.

ii. ***Biennials:*** Biennials are those crops which complete their life cycle in two years. There is vegetative growth in the first year while there is reproductive stage in second year. e.g. Sugarbeet etc.

iii. ***Perennials:*** These crops takes more than two years to complete their life cycle. e.g. Lucerne, cotton, tea, different grasses etc.

B. According to Seasons

i. ***Rainy or Kharif season crops:*** Crops require high temperature and high humidity at the time of sowing and harvesting. Sowing is done in June–July. e.g. Rice, sorghum, bajra, maize, moong, urd, cowpea, groundnut etc.

ii. ***Winter or Rabi season crops:*** Crops require low temperature at sowing while hot and dry weather at the time of harvesting. Sowing is done in October – November. e.g. Wheat, gram, barley, lentil.

iii. ***Spring or summer or zaid season:*** These crops are short duration crops and matures in 3 months. e.g. Urd, moong, cowpea, maize, cheena, cucumber, muskmelon, watermelon.

C. According to their Family : (*Botanical classification*)

i. ***Gramineae or Poaceae or grass family*:** e.g. wheat, rice, jowar, bajra, sugarcane, barley, oat etc.

ii. ***Leguminaceae family*:** e.g. pea, gram, fababean, lucerne, sunhemp, cowpea, berseem etc.

iii. ***Cruciferae family*:** e.g. mustard, til, cauliflower, safflower, rai, lahi, tarameera etc.

iv. ***Solanaceae family*:** eg. potato, tobacco etc.

v. ***Tiliaceae family*:** e.g. jute

vi. ***Liliaceae family*:** e.g. linseed

vii. ***Euphorbiaceae family*:** e.g. castor (Arandi)

viii. ***Compositeae family*:** e.g. sunflower

ix. ***Chenopodaceae family*:** e.g. sugarbeet

x. ***Malvaceae family*:** e.g. cotton, patsan etc.

D. According to Agronomic/Economic basis

i. ***Cereal crops*:** Wheat, maize, paddy, barley, oat, pearl millets, finger millets, foxtail millet, sorghum, maize.

ii. ***Legumes and Pulses*:** They are rich source of protein. eg. moong, gram, pigeon pea, green gram, black gram, soybean, peas, cowpea, horse-gram, lentil.

iii. ***Forage crops***: These crops are used for feeding animals e.g. Sorghum, maize, berseem, oats, lucerne, elephant grass, guinea grass.

iv. ***Fibre crops***: These crops are used as fibres e.g. Cotton, flax, sunhemp, jute, mesta.

v. ***Sugar crops***: These crops are used for sugar purpose e.g. Sugarcane, sugarbeet etc.

vi. ***Oil seed crops***: These crops are used for oils. e.g. Lahi, arandi, groundnut, linseed, sunflower etc.

vii. ***Stimulant and Medicinal crops***

   i ***Stimulant***: These crops stimulates the process. e.g. Tobacco, tea, coffee, etc.

   ii ***Medicinal***: These crops have medicinal properties. e.g. Mentha, citronella, lemon grass, jasmine, geranium, cassava, elephant yam etc.

viii. ***Root and Tuber crops***: In these crops, food material is stored in underground root and stem. eg. Potato, carrot, sweet-potato, radish, sugarbeet, onion, etc.

ix. ***Spices crops***: Ginger, garlic, onion, coriander, turmeric, ginger, etc.

x. ***Vegetable crop***: e.g. Radish, pumkin, carrot, okra, brinjal, tomato etc.

xi. ***Fruit crops***: e.g. Mango, litchi, guava, apple, papaya, apricot etc.

E. According to Special use

i. ***Green manuring crops***: Crops, which are used for manuring. They produce green vegetative growth and are turned and mixed into the soil at the stage of pre-flowering. e.g. Sunhemp, dhaincha, lobia, moong, urd, guar etc.

ii. ***Cover crops*** : These crops have vigorous vegetative growth which covers the upper surface of the soil to protect the soil erosion from air and water. Then, after this, they are turned into green manure. e.g. Soybean, cowpea, moong, urd etc.

iii. ***Catch crops***: Crops which are grown between two main crops and mature early. e.g. Lahi, moong, cheena etc. These crops are also grown when main crop fails.

iv. ***Cash crops***: Cash crops are not consumed as a whole by the farmers family it self, and extra produce is sold in the market to earn money. eg. Sugarcane, tobacco, soybean, potato, cotton.

v. ***Trap crops***: The crops, which are grown on the border of the field to protect the main crop from damage (insect, pest). These crops have the ability to attract the insects. e.g. Bhindi is grown as trap crop around the cotton crop and sunflower around maize crop.

vi. ***Arable crops***: These requires preparatory tillage. e.g. Wheat, maize etc.

vii. ***Mulch crops***: To conserve soil moisture. eg. Leguminous crops.

viii. ***Smoother crops***: They are used to supress the population and growth of weeds by providing dense foliage and quick growing ability. e.g. Cowpea, mustard etc.

F. Simple Classification

i. ***Field crops***: Cereals, millets and pulses.

ii. ***Plantation crops***: Tea, coffee, rubber, coconut.

iii. ***Commercial crops***: Oilseed crops, sugar crops, fibre crops and tobacco.

iv. ***Horticultural crops***: Fruits, vegetable and ornamentals.

v. ***Forage and grasses***: Sorghum, maize, elephant grass, guinea grass and other pulse crops.

vi. ***Condiments and spices***: Cardamom, pepper, chillies, ginger and turmeric.

vii. ***Medicinal and aromatic plants***: Same as under agronomic classification.

Chapter 3

Soil: Origin, Classification and its Properties

Soil and its Components

Soil

Soil is a three dimensional, transferable natural material which is found on the earth crust and provides natural medium to plant growth. OR

Soil is the upper most loose layer of the earth, suitable for plant growth.

Different scientists defined 'SOIL' in different ways:

Buckman and Brady defined soil as "A dynamic natural body on the surface of the earth in which plants grow, composed of mineral and organic materials and living forms".

Raman defined soil as the uppermost weathered layer of the soil earth's crust; it consists of rocks that have been reduced to small fragments and have been more or less changed chemically together with the remains of plants and animals that lived on it and in it."

According to USDA , "Soil is a natural body comprised of solids (minerals and organic matter), liquid, and gases that occurs on the land surface, occupies space, and is characterized by one or both of the following: horizons, or layers, that are distinguishable from the initial material as a result of additions, losses, transfers, and transformations of energy and matter or the ability to support rooted plants in a natural environment."

Functions of soil

- It provides mechanical support.
- It regulates water supplies.
- It supply essential mineral nutrients.
- It supply air for better growth of roots and other microbes present in the soil.
- It provides habitat for soil organisms.

Soil Composition

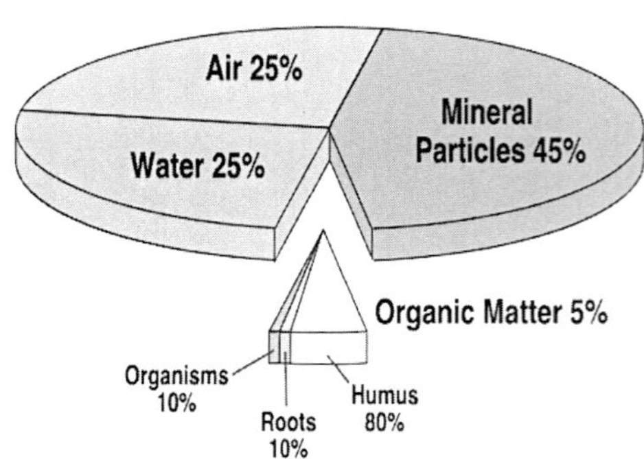

Figure : Most soils contain four basic components: mineral particles, water, air, and organic matter. Organic matter can be further sub-divided into humus, roots, and living organisms. The values given above are for an average soil

Soil consists of four major components

1. ***Mineral matter (inorganic form)*:** It constitutes about 45 percent. There are two types of minerals present in soils.

Primary minerals: These are similar to parent material and present in the coarse section of the soil. eg. Quartz, feldspar, mica, calcite, biotite.

Secondary minerals: These are derived from chemical weathering of rocks.These are found in fine section of soil. e.g. Dolomite, gypsum, apatite, limonite, silicate, clay, limonite, haematite.

2. ***Organic matter*:** It constitute about 5 percent. It is partially decayed and partially synthesized plant and animal residues. Organic matter can be further divided into humus, roots and organisms.

Humus: It constitutes 80% of organic matter. It is the biochemical substance that makes the upper layer of the soil to become dark.

Roots: It constitutes 10% of organic matter. Roots of different plants are also the sources of soil organic matter.

Soil organisms: It constitutes 10% of organic matter and further divided into macroflora and microflora.

3. ***Water***: It constitute about 25-30 percent. It acts as medium for plant growth, as a solvent and carrier of nutrients. It is involved in different chemical, physical and biological activities in the soil. It maintains turgity of plants and helps in photosynthesis. It acts as an agent in weathering of rocks and minerals.

4. ***Air***: It constitute about 25-30 percent. It is the gaseous phase of soil which consists of gases like nitrogen, oxygen, carbon dioxide and water vapour. It is a path way through which oxygen is taken in and absorbed by soil micro organisms and plant roots and carbon dioxide produced by the plants is removed. This two way process is called soil aeration.

Macroorganisms like insects, rodents and worms and microorganisms like bacteria, fungi and algae are also found in the soil. The relative proportions of these four soil components vary with soil type and climatic conditions.

Soil Profile

Soil profile is the vertical section of soil upto the level of parent material (bed rock). If a section of soil is cut downward, we will find different layers one over another, such a section is called profile, and the individual layer is called as horizons. They range from rich, organic upper layers (humus and topsoil) to underlying rocky layers (subsoil, regolith and bedrock). These horizons above the parent material are collectively referred to as the solum (Solum is derived from Latin word, means soil or land). Every well developed, undisturbed soil has its own distinctive profile characteristics. They are useful in soil classification, survey and in judging soils of its importance.

Soil consists of four major zones (horizons)

1. ***O horizon*** – The top, organic matter layer, which is productive (Aoo & Ao) and mostly made up of leaf litter and humus.

2. ***A horizon*** (**Zone of leaching** or **eluviations**): It is also called mineral soil layer and also called top soil. Cations are leached from this horizon by strongly acid solutions generated in the O horizon.

3. ***B horizon*** (**Zone of accumulation**)**:** It is also called sub soil or the zone of deposits where the cations leached out of the A horizon accumulates. Horizon consists of clays, iron and aluminum oxides.

4. ***C horizon***: It is also called Regolith or zone of parent materials or unweathered materials. It is the lower most zone.

Below this is the bed rock

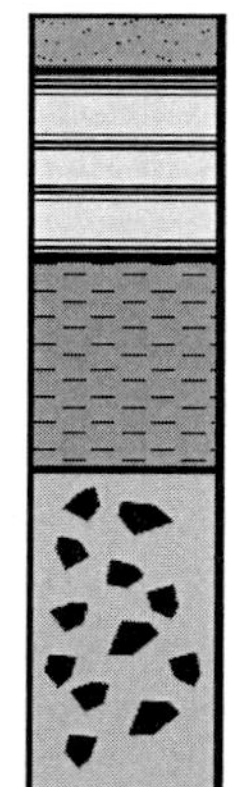

Difference between Soil and Sub-soil

Soil	Sub-soil
1. It is dark in colour	1. It is light in colour
2. Organic matter and nitrogen is high	2. Organic matter is low and potassium content is high
3. The soil is completely weathered	3. The sub- soil is less weathered
4. Generally all the nutrients are available	4. Generally all the nutrients are not available
5. Soil is fertile	5. The sub-soil is less fertile
6. Soil is fine textured	6. Sub-soil is coarse textured
7. The soil is loose and have more aeration	7. The sub- soil is compact and have poor aeration
8. The soil has more number of microorganisms	8. The soil has less number of microorganisms

Soil Formation and Soil Forming Factors

Soil formation

Soils are formed by the mechanical disintegration and chemical decomposition of parent rocks. The loose material after various weathering processes, develops a soil profile, with different horizons and addition of organic materials etc. is called soil development. Soil formation is a slow process. Formation of one inch soil needs 800-1000 years.

The soil formation is the process of two consecutive stages.

1. The weathering of rock (R) into regolith.
2. The formation of true soil from regolith.

The evolution of true soil from regolith takes place by the combined action of soil forming factors and processes.

1. The first step is accomplished by weathering (disintegration and decomposition).
2. The second step is associated with the action of soil forming factors.

Weathering of Rocks: Weathering is a process whereby the solid rocks of the earth's crust are broken down to form the parent material of soil. Weathering is the process which acts at the earth's surface to decompose and breakdown rocks.

Types of Weathering: There are three types of weathering.

A. Physical or Mechanical Weathering : It is the breakdown of rock material into smaller and smaller pieces with no change in the chemical composition of the weathered material. Weathering takes place by the action of physical agencies like water, wind, temperature, pressure, plants and animals. Soils formed as a result of physical weathering are called *skeletal soils*. In mechanical weathering following processes takes place.

Expansion and contraction : Thermal heating and cooling of rocks causing expansion and contraction.

Freezing and thawing: Water freezes at night and expands because the solid occupies greater volume, breaks off the fragments and displaces masses. Thawing results rock fragments to roll down the slopes.

Exfoliation: Process in which curved plates of rock are stripped from a larger rock mass.

Grinding and rubbing: The grinding and rubbing of moving rocks against each other by water, wind, glaciers result in very effective disintegration.

Other types: Cracking of rocks by plant roots and burrowing of animals.

Soil in early stages of development and the soils of arctic, alpine and deserts are formed mainly due to the result of physical weathering.

B. Chemical Weathering: It is simply the breakdown of rocks by chemical agents. It involves two phases:

i. Disappearance of certain materials.
ii. Formation of secondary products.

Chemical weathering is less under desert conditions due to scarcity of water while under arctic and alpine conditions due to low temperature.

The chemical decomposition of rocks is brought about by the followings chemical process.

Solution: Most minerals are soluble in water to some extent and so chemical reactions are setup in the soil which result in the formation of new substance. Rocks contains most of the soluble salts i.e. gypsum ($CaSO_4$), Limestone ($CaCO_3$) dissolve easily in water.

$$H_2O + CO_2 \rightarrow H_2\,CO_3$$

Carbonic acid has solubilizing effect

Hydration: It implies the association of water molecules with minerals. Mineral compounds absorb water and destruct the chemical structure.

$$2\,Fe_2\,O_3 + 3H_2O \rightarrow 2\,Fe\,O_3.\,3H_2\,O$$

Hematite Limonite

$$KAl.Si_3O_8 + H_2O \rightarrow H.Al.Si_3O_8 + KOH$$

Orthoclase Acid silicate (soluble)

Hydrolysis: In the process of hydrolysis, minerals split by water and form hydroxide of new, more soluble substances. During hydrolysis, desilcification, dealkalization also occur simultaneously with the formation of new compounds.

$$K_2O.Al_2O_3.6SiO_2 + 2\,H_2O \rightarrow H_2O\,Al_2O_3.6SiO_2 + 2KOH$$

Feldspar

$$H_2O\,Al_2O_3.6SiO_2 + H_2O \rightarrow Al_2O_3.2SiO_2 + 2\,H_2O + 4SiO_2$$

Kaolinite Silica

$$2H.Al.Si_3O_8 + 8\,H_2O \rightarrow Al_2O_3.3H_2O + 6SiO_2.\,H_2O$$

Acid aluminosilicate Hydrated alumina Hydrated silica (soluble)

Due to presence of KOH, the weathering will be faster.

Carbonation : The hydroxides produced during hydrolysis react with the dissolved CO_2 to form corresponding carbonates which may either leach out or accumulated according to drainage and weather conditions.

$$CO_2 + H_2\,O \rightarrow H_2\,CO_3$$

$$Ca\,CO_3 + H_2\,CO_3 \rightarrow Ca\,(HCO_3)_2$$

Calcite (Less soluble) Carbonic acid (Readily soluble)

$K.Al.Si_3O_8 + H_2CO_3 \rightarrow H_4Al_2Si_2O_9 + K_2CO_3 + 4SiO_2$

Feldspar Carbonic acid Aluminium silicate Silica (soluble)

Oxidation: Oxidation is the addition reaction of oxygen upon minerals. Oxygen is more active in the presence of moisture and resulted in hydrated oxides. Rocks containing sulphites, carbonates and silicates of iron undergo oxidation.

$$4FeO + O_2 \rightarrow 2Fe_2O_3 \rightarrow 2Fe_2O_3.3H_2O$$

Ferrous oxide Ferric oxide Limonite

$$4FeCO_3 + O_2 \rightarrow 2Fe_2O_3 + 4CO_2$$

Reduction: Under unaerobic, reduced condition, oxygen is removed out due to lack of oxygen

$$2Fe_2O_3 \rightarrow O_2\uparrow + 4FeO$$

Ferric oxide

Deposition: It occurs usually in the belt of cementation. When Iron (Fe) and aluminium (Al) moves as colloids and comes in contact with electrolytes, it gets deposited due to intermitted heating and drying.

C. Biological Weathering: Joffe defined Biological weathering as "Essentially it is physical and chemical weathering by biological agencies". Biological agents such as vegetation, animals and human action increase both types of weathering i.e. physical and chemical. Plant roots helps in widening of cracks and crevices in the rocks. Burrowing animals is very useful in biological weathering by exposing the sub soil to weathering. Lower forms of plant life like mosses and lichens first appear on the broken and physically comminuted rock fragments.

Soil Forming Factors

Dokuchaiev (1889) established that the soils develop as a result of the combined action of soil forming factors

S = f (cl,o,r.p,t,.......)

S = Soil formation

cl = climate

o = Organisms and vegetation (biosphere)

r = Relief or topography

p = Parent material

t = Time

.. - additional unspecified factors

According to Jenny " Soil property is determined by the relative influences of these factors" The five soil forming factors, acting simultaneously at any point on the surface of the earth, to produce soil.

Parent material: Rocks, minerals and their nature.

Climate: Heat, water, pressure, wind, air, etc.

Topography: It influences the soil formation by slopes, high and low altitudes.

Biosphere: It includes plants, animals, micro-organisms, bacteria, fungi, wild life, animals and humans affect soil formation.

Time: Time for all these factors to interact with the soil is also a factor. Over time, soils exhibit features that reflect the other forming factors.

Rocks: Rocks are chief soil forming parent materials over which the soils are developed.

Classification of Rocks

A. On the basis of silica content, rocks are classified as

- **Acid rocks**: Silica content is more than 65-75% i.e. granite and sand stone.
- **Intermediate rocks**: Silica content is 55-65% i.e. syenite, diorite, andecite.
- **Basic rocks** : When silica content is less than 55% then the nature of rock is basic *i.e* Gabbro, basalt, lime stone, dibase. These are rich in Fe, Ca, Mg and Na.

The rocks having complex mineral composition weather more easily such as basic rocks weathered at a faster rate than the acid rocks. Generally, acid igneous rocks produce soils with good physical conditions but poor in chemical properties.

B. On the basis of genesis and structure, rocks are classified as

- **Igneous rocks**: They are formed from the solidification of molten mass of earth (magma). These rocks are acidic in nature. Igneous rocks are of two types:

  a. ***Plutonic or Intrusive rocks***: These rocks are formed inside the earth by sedimentation or solidification of the magma and pushed to the surface by earth movements in pre-geological periods. The formation slow due to slow cooling processes. During cooling, crystals developed and the rocks that formed were crystalline. i.e. granite, diorite, syenite, gabbro.

b. ***Volcanic or Extrusive or Effusive rocks**:* In these types of rocks, cooling of lava takes place outside on earth surface and the process of solidification is rapid. Structure is glassy i.e. Basalt, Andesite, Trachyte.

- **Sedimentary or Derivative rocks:** With the changes in temperature, the action of ice, heat and cold and gigantic movements of earths, igneous rock has broken down into smaller pieces which have been transported and deposited to form new rock masses called sedimentary or derivative rock. These rocks cover $4/5^{th}$ of the extent of rocks visible on the surface of the earth, yet, they may not constitute more than 50% of the total bulk, the igneous rocks make up the rest. These are basic in nature because they contain calcium, sodium, potassium. e.g. Limestone, quartzites, dolomite, shales, sandstone, conglomerate.
- **Metamorphic rocks**: These rocks resulted from the subsequent transformation of igneous rocks and sedimentary rocks under the influence of intense heat, pressure and chemical action of gases and liquid substances. eg. Gneiss from Granite, Slate from Shale, Marble from Lime stone, Quartz from Sandstone, Quartzite from Quartz or Sandstone and Scyst from Shale.

Properties of Soil

There are three properties of soil:

A. Physical properties of soil

B. Chemical properties of soil

C. Biological properties of soil

A. Physical Properties of Soil

Physical properties (mechanical behaviour) of a soil greatly influence its use and behaviour towards plant growth. The plant support, root penetration, drainage, aeration, retention of moisture, and plant nutrients are linked with the physical condition of the soil. Physical properties also influence the chemical and biological behaviour of soil. The physical properties of a soil depend on the amount, size, shape, arrangement and mineral composition of its particles. These properties also depend on organic matter content and pore spaces.

1. Soil colour: Soil colour may be inherited from its parent material. It depend upon moisture content, organic matter content, colloidal clay and parent material. Soil colour is also very useful in identification and description of soil groups i.e. clay has dark colour but sandy soil has light colour. We can judge the agricultural value of soil by looking soil colour. Black and dark colour of soil is mainly due to organic matter content and iron, whereas white colour is due to silica and lime.White colour is common due to salt or carbonate deposits.

2. Soil density: Soil density is expressed as mass of soil per unit volume. Its unit is gram per cubic metre. It depends on texture, structure and presence of organic matter.

Soil density is expressed in two well- accepted concepts

- **Bulk density (Bd) or Apparent specific gravity :** The oven dry weight of a unit volume of soil, which includes pore spaces, organic material is called bulk density. Generally, soils with low bulk densities have favourable physical conditions and vice-versa. Fine textured soils like clay, silt loam etc. have lower Bd than sandy soil due to high pore spaces. Soil having bulk density between 1.4-1.6 g/cc is considered good for the plant growth.

  Bulk density (Bd)= weight of soil/volume of solids and pores.

- **Particle density (Pd) or Real specific gravity**: It is the weight per unit volume of the solid portion of the soil. Generally particle density of normal soil is about 2.65 g/cc. Heavy mineral containing soils (i.e. magnetite, Limonite, hematite) have greater particle density, than soil containing high organic matter. It is also termed as true density.

Texture class	P.density (g/cc.)	B.density (g/cc.)	Pore space(%)
Sandy	2.65	1.6	40
Silt	2.80	1.3	50
Clay	2.94	1.1	60

3. Soil porosity : It refers to that percentage of soil volume which is occupied by pore spaces. Pore space is the volume of soil mass that is not occupied by soil particles is known as pore space. It is occupied by air and water and plant roots grow and developed in these spaces. It helps in judging the moisture movement within the soil. Soil porosity can be calculated by the formula:

Porosity = 100 - (B.density/ P.density) x 100

Since, % pore space + % solid space = 100

% pore space = 100 - % solid space

% Solid space = (B. density / P. density) x 100

Porosity or (% pore space) = 100 – (Bd / Pd) x 100

There are two types of pores in the soil-

a. **Macro or non-capillary pores**: These are the large pores, movement of air and water is easy. Sand and sandy soil have a larger number of macro pores. Such soils do not held water much.

b. **Micro or capillary pores:** The movement of air and water is restricted to some extent. Clays and clayey soils have a greater number of micro pores. In all, total pore space is more in clay (heavy) soils than light sandy.

For better crop growth and development, the both pore-spaces should be in an equal proportion. It provides an ideal condition for aeration, permeability, drainage and water retention.

4. Soil texture: It refers to the relative proportion of soil particles of various size or relative percentage of sand, silt and clay in a soil.

Soil Textural classes: Soil particles are grouped into following classes:

Sand: The soil containing more than 70% of sand seperates are called sands. They have irregular shape, greater pore space, lower water retention capacity maximum leaching and lesser adsorption surfaces for nutrients.

Clay: The soil containing 40% or more of clay particles are classed as clay soil. These are less porous, more sticky, have more water retention capacity, absorb more nutrients. These soils are further divided into sandy clay, silty clay, clay, fine clay etc.

Loams: The soil consisting mixtures of seperates without domination of any group is called loams. The pore spaces, leaching, water holding capacity lies between sands and clay soils. These have good physical character. They are divided into sandy loam, silt loam, clay loam etc.

There are different systems of naming soil separates.

Table: Classification of soil particles (International system)

Soil separates	Dimeter range (mm)
Stone	More than 20
Gravel	2.00-20
Fine earth	Less than 2.0
Coarse sand	0.20-2.0
Fine sand	0.02-0.2
Silt	0.002-.02
Clay	Less than 0.002

Table: Textural classes by U.S.D.A.

Common name	Texture Basic soil	Textural class name
Sandy soils	Coarse	Sandy, Loamy sands
Loam soils	Moderately coarse	Sandy loam, Fine sandy loam
	Medium	Very fine sandy loam, Loam, Silt loam, Silt
	Moderately fine	Clay loam, Sandy clay, Loam, Silty clay loam
Clayey soils	Fine	Sandy clay, Silty clay, Clay

5. Soil structure : The arrangement of soil particles or aggregate into certain defined patterns is called soil structure.The primary soil particles, i.e. sand , silt, clay are grouped together to form aggregates. Naturally occurring aggregates are called Peds, while artificially formed soil mass are Clods. Structure is studied in the field under natural conditions and described under three categories.

i. **Type :** Shape or form and arrangement pettern of peds.

ii. **Class :** Size of peds.

iii. **Grade :** Degree of distinctness of peds.

Types of structures

Plate or plate like: Aggregates are arranged in relatively thin horizontal plates. The horizontal dimensions are much developed than the vertical giving a flattened, compressed or lens like appearance to the peds. When the units are thick then they are called platy and when the units are thin they are called laminar plate. Platy structure is commonly found in sub soil surface of virgin soil and often inherited from the parent material laid down by water.

Prism like or prismatic: The vertical axis is more developed than horizontal, giving a pillar like shape. When the top of such a ped is rounded, the structure is termed as columnar, and when flate then it is known as Prismatic such structure of peds occurs in sub soil horizons in arid and semi –arid regions.

Block or Block like: The dimensions in block- like structure are almost similar in size and the peds are cube like with flat or rounded faces. When the faces are flat and the edges are sharp angular, the structure is named as angular blocky. When the faces and edges are round in shape it is called sub - angular blocky. Blocky type structure is found in sub soil.

Spheroidal or sphere like: All the rounded aggregates come under this category. These rounded peds are lying loosely or separately. On wetting, the spheres do not collapse and remain porous after swelling. These are of two types.

i. ***Granular*:** In case of spheriodal structures, the aggregates are rounded but when curved or irregular and less porous, the structure is called granular.

ii. ***Crumby*:** Spheroidal structure, the aggregates are rounded and very porous, it is called crumb structure. Amongst all kind of structures described above the crumb structure is the most ideal for crops.

Grades of structure

Grade indicates the degree of distinctness of the individual peds. Four terms commonly used to describe the grade of the soil structure are:

1. ***Structureless:*** There are no noticeable peds such as conditions exhibited by loose sand or cement like conditions of some clay soils.
2. ***Weak:*** Poor formation of peds, which are barely durable.
3. ***Moderate:*** Moderately well developed peds, which are fairly durable and distinct.
4. ***Strong:*** Strong well developed peds are durable and distinct.

6. Soil consistence : It is a dynamic properties of soil. It refers the behavior of soil in its degree and kind of cohesion and adhesion or its resistance to deform or rupture.

Cohesion refers to the attraction of substances of like characteristics such as that of one water molecule for another.

Adhesion is the attraction of unlike materials *i.e.,* attraction between soil and water molecules.

Phenomena of consistency are :

Friability : The soil is said to be friable when soil clods are easily crushed into smaller aggregates. The soil is loose, non-coherent. Soil water is at optimum enough to fill the soil (most field operations are done).

Plasticity : Plasticity is the capacity to be molded (toughness). When stress is applied to wet soil, the shape is changed and after removing stress, the shape is remain unchanged. The terms used to described the degree of plasticity are: non plastic, slightly plastic, plastic or very plastic.

Stickness : The quality of adhesion to other objects is called stickness. The degree of stickness *i.e.,* non – sticky, slightly sticky, sticky, very sticky.

Hardness : When soil is dry, the soil consistence is characterized as hardness or rigidity. It is described *i.e.* loose – non – coherent, soft – breaks under slight pressure between thumb and become powdery, hard – breaks tightly, very hard – do not break with thumb.

7. Soil temperature: Plant growth, along with physical and biological activities in the soil are greatly affected by the soil temperature. It provides suitable medium for optimum growth, activity of microorganisms, nutrient movement, germination, water absorption by plants etc.

8. Soil air: It is the gaseous phase of soil which consists of gases like nitrogen, oxygen, carbon dioxide and water vapour. Ordinarily, occupation of nearly $1/3^{rd}$ of the pore space in the soil by air and $2/3^{rd}$ by water constitutes the most favourable condition for plant growth.

Table: Percentage by Volume

	Oxygen (O_2)	Carbon-di-oxide(CO_2)	Nitrogen(N_2)
Soil Air	20.6	0.25-1.0	79.2
Atmosphere	20.96	0.03	79.0

Soil air in sandy soil is > (greater or equal to) 25 %, in loamy soil is 15-20% and in clay soil <10 % of the total soil volume. Clay soil retain more water hence is lower air capacity.

9. Soil water: It act as medium for plant growth, as a solvent and carrier of nutrients. It is involved in different chemical, physical and biological activities in the soil.

Classification of Soil Water

Soil water has been classified into Physical classification of soil water, and Biological classification of soil water.

A. Physical classification of soil water

1. Gravitational water: Gravitational water occupies the larger soil pores (macro pores) and moves down readily under the force of gravity. Water in excess of the field capacity is termed gravitational water. Gravitational water is of no use to plants because it occupies the larger pores. It reduces aeration in the soil. Thus, its removal from soil is a requisite for optimum plant growth. Soil moisture tension at gravitational state is zero or less than 1/3 atmosphere.

2. Capillary water: Capillary water is held in the capillary pores (micro pores). Capillary water is retained on the soil particles by surface forces. It is held so strongly that gravity cannot remove it from the soil particles. The molecules of capillary water are free and mobile and are present in a liquid state. Due to this reason, it evaporates easily at ordinary temperature though it is held firmly by the soil particle; plant roots are able to absorb it. Capillary water is, therefore, known as available water. The capillary water is held between 1/3 and 31 atmosphere pressure.

3. Hygroscopic water: The water that held tightly on the surface of soil colloidal particle is known as hygroscopic water. It is essentially non-liquid and moves primarily in the vapour form. Hygroscopic water held so tenaciously (31 to 1000 atmospheres) by soil particles that plants can not absorb it and water moves in vapour form. Some microorganism may utilize hygroscopic water.

Biological Classification of Soil Water

There is a definite relationship between moisture retention and its utilization by plants. This classification based on the availability of water to the plant. Soil moisture can be divided into three parts.

1. Available water: The water which lies between wilting coefficient (15 atm) and field capacity (1/3 atm). It is obtained by subtracting wilting coefficient from moisture equivalent.

2. Unavailable water: This includes the whole of the hygroscopic water plus a part of the capillary water below the wilting point. Such water is held in soil at the permanent wilting point (PWP).

3. Super available or superfluous water: The water beyond the field capacity stage is said to be super available. It includes gravitational water plus a part of the capillary water removed from larger interferences. This water is unavailable for the use of plants. The presence of super-available water in a soil for any extended period is harmful to plant growth because of the lack of air.

B. Chemical Properties of Soil

These are pH of soil, cation exchange capacity; buffering capacity and soil colloids. These are having more significance in the crop production.

Soil pH

Soil pH is a measure of the acidity or alkalinity in the soil. It is also called soil reaction. Soil solution contains different salts in ionized state. If soil solution contains equal proportion of acidic and basic ions, the reaction is said to be neutral. If there are more H^+ ions in the soil solution, then the soil is acidic in reaction while if soil contains basic ions like Ca^{++}, Mg^{++}. Na^{++}, then the soil is basic. Maximum availability of nutrients lie between pH 6.5 to 7.5.

The most common classes of soil pH are:

Extremely acid 3.5 – 4.4
Very strongly acid 4.5 – 5.0
Strongly acid 5.1 – 5.5
Moderately acid 5.6 – 6.0
Slightly acid 6.1 – 6.5
Neutral 6.6 – 7.3
Slightly alkaline 7.4 – 7.8
Moderately alkaline 7.9 – 8.4
Strongly alkaline 8.5 – 9.0

Cation Exchange Capacity (CEC): It is the sum of total cations that soil can absorb. If CEC = 10 c mol/kg, it means 1 kg of soil can absorb 10 c mol of H^+ and can exchange with 10 c mol of monovalant cations (K^+, Na^+) or 5 c mol of divalent cations (Ca^{2+}, Mg^{2+}). CEC is influenced by soil pH.

Buffering capacity: It is the ability to resist the change in soil pH

Soil colloids: Soil contains two types of colloids:

i. Mineral or inorganic colloids which include clay minerals

ii. Organic colloids which include humus

C. Biological Properties of Soil

Soil is not a dead mass but an abode of millions of organisms, which includes crabs, snails, earthworms, mites, millipedes, centipedes. These feed on plant residues burrow the soil and help in aeration and percolation of water.

The soil organisms are of two types: Microflora and Microfauna

Microflora : Algae, Fungi, Actinomycetes Bacteria.

Microfauna: Protozoa and nematodes

Classification of Soil or Soil groups

Indian soils have been classified into different groups and sub-groups. According to United States Department of Agriculture USA, nomenclature system, soils are classified as shown in Table.

Table . The extent and distribution of the different soil classes of India as represented in the soil maps on 1 : 250,000 scale along with their equivalent according to United States Department of Agriculture USA, nomenclature system

S. No.	Major soil (traditional name)	Extent '000 ha	Percentage	Distribution in states US soil taxonomy	Soil orders
1.	Alluvial	100,006	30.4	J&K, HP, Punjab, Haryana, Delhi, UP, Gujarat, Goa, MP, MS, AP, Karnataka, TN, Kerala, Puducherry, Bihar, Odisha, WB,ArP, Assam, Nagaland, Manipur, Mizoram,Tripura, Meghalaya, A&N	Inceptisols, Entisols, Alfisols, Aridisols
2.	Coastal alluvial	10,049	3.1	AP, Karnataka, TN, Kerala, WB, Gujarat, Odisha, Puducherry, Lakshadweep, A&N	Aridisols,Inceptisols, Entisols
3.	Red	87,989	26.8	AP, Karnataka, Kerala, TN, Puducherry, Rajasthan, MP, MS, Gujarat, Goa, ArP, Assam, Manipur, Meghalaya, Nagaland, Mizoram, Tripura, Delhi, UP, HP, A&N	Alfisols,Ultisols, Entisols, Inceptisols, Mollisols, Aridisols
4.	Laterites	18,094	5.5	AP, Karnataka, Kerala, TN, Puducherry, MS, Odisha, WB	Alfisols, Ultisols, Entisols
5.	Brown forest	540	0.2	Karnataka, Maharashtra	Mollisols, Inceptisols
6.	Hill	2,262	0.7	Manipur, Odisha, WB, Tripura, Nagaland	Inceptisols, Entisols
7.	Terai	326	0.1	UP, Sikkim	Mollisols, Entisols
8.	Mountain meadow	60	—	J&K	Mollisols
9.	Sub-montane	104	—	J&K	Alfisols
10.	Black	54,682	16.6	MP, MS, Rajasthan, Puducherry, TN, UP, Bihar, Odisha, AP, Gujarat	Vertisols, Mollisols, Inceptisols Entisols, Aridisols
11.	Desert	26,283	8.0	Rajasthan, Gujarat, Haryana, Punjab	Aridisols, Inceptisols, Entisols
	Others*	28,305	8.6	—	—
	Total	328,700	100	—	—

*Includes glaciers (0.4%), sand dunes (0.01%), mangrove swamps (0.04%), salt waste (0.01%), water bodies (0.1%), rock land (0.25%) and rock outcrops (7.8%). MP, Madhya Pradesh; MS, Maharashtra; UP, Uttar Pradesh; J&K, Jammu and Kashmir; TN, Tamil Nadu; AP, Andhra Pradesh; ArP, Arunachal Pradesh; WB, West Bengal; HP, Himachal Pradesh; A&N, Andaman and Nicobar Islands.

Source: *Govinda Rajan, S. V., Soil Map of India. In Review of Soil Research in India (eds Kanwar, J. S. and Raychaudhuri, S. P.), 1971, 1:7.*

Different soils

1. Alluvial Soils (Fluvisols): They are also called as Indo – gangrenous alluvium. Alluvial soils constitute the largest and most important soil group of India. The soils are derived from the deposition of silt by three great Himalayan rivers i.e. Sutlej, Ganga and Brahmaputra and its tributaries. Alluvial soils, are divided into

a. Coastal Alluvium b. Coastal Sands c. Deltaic Alluvium
d. Calcareous Alluvial Soils

Soils: Soils are deep, sandy loam to clay loam, well drained and yellow brown to dark brown in colour. These are the most important soils from the agriculture point of view. Most of the alluvial soils have been classified in the orders, Entisols, Inceptisols and Alfisols.

Nutrient status: Alluvial soils are generally deficient in Nitrogen, humus and occasionally in rich in potassium. Alluvial soils have low base exchange capacity.

pH: Nearly neutral to slightly alkaline.

Crops suitable under alluvial soils: Wheat, cotton, maize, millets, rice, oil seeds, Sugarcane, Vegetables and fruits.

2.Coastal Alluvial: Its texture is sandy to silty clay. The soils are usually deep and its colour varies from bright reddish brown to yellowish brown and to grey and dark grey. These soils are derived from calcareous material and are composed of dark coloured heavy clay. And if they are derived from red soils which are derived from granite and gneiss rocks, the soils are poor in fertility with medium or light texture and rich in kaolinitic clays.

a. **Coastal sands:** These soils are sandy and deep but lack in profile development. Coconut and cashew plantations are common in these soils.

b. **Deltaic alluvium:** These soils are brought by rivers and deposited where they join the sea.

c. **Calcareous alluvial soils:** These are calcareous soils developed on the Alluvium of Gandak river. These soils are high in calcium carbonate (10-40%) and its texture varies from sandy loam to loam and pH is alkaline.

3. Red soils (Regur soils)

Soil: These soils have been formed due to weathering of ancient crystalline and metamorphic rocks. These soils are light textured with porous structure. Lime is absent with low soluble salts. Most of the red soils have been classified in the order 'Alfisols'.

Nutrient status: Red colour is due to various oxides of iron. They are deficient in nitrogen, potassium, phosphorus and humus.

pH: 7-7.5

Crops suitable under red soils: Millets, pulses, linseed, tobacco etc.

4. Laterite and lateritic soils

Soil: Word laterite is derived from latin word 'laterite' meaning brick. Laterite soils are formed by weathering of lateritic rocks, low temperature and heavy rainfall with alternating dry and wet periods. These soils are present in high rainfall areas and under high rainfall conditions, silica is released and leached downwards and the upper horizons of soils become rich in oxides of iron and aluminium. The texture is light with free drainage structure. Laterite soil does not retain moisture and hence is not fertile. Most of the laterite soils have been classified in the order ' ultisols' and a few under 'oxisols'.

These are of two types: Upland laterites and Lowland laterites

Upland laterite: are formed over hills and uplands. From uplands, these are transported by streams towards lowlands. Such transported soils are known as Lowland laterites.

Nutrient status: These soils have fairly high organic matter content but poor in N, P, K, Mn and lime. These soils are less fertile.

Crops suitable under laterite soils: Rice, sugarcane, tea, rubber, coffee, tapioca, cashewnut.

5. Hill soil (Brown) : These soils are formed from sandstone, grey micaous sandstones and shales. The surface soils are brown in colour and moderately rich in organic matter. Its texture is loam to silty clay. They have high base exchange capacity.

pH: neutral to acidic

Crops suitable under Hill soil: Rice, wheat, millets, vegetables, peach, plum etc.

6. Terai soils: These soils are formed by the downward movement of materials from the lower Himalayan ranges. Surface soils are sandy loam or silty loam in texture and with the efficient drainage, these become fertile soils. These soils are found in the foot of Himalayas, Jammu and Kashmir, Bihar and West Bengal.

7. Sub-montane: Found under coniferous type vegetation and has large accumulation of organic matter. Lime is absent in these soils.

pH: Neutral to acidic

8. Black cotton soil: These soils are also known as Vertisols or Regur soil or Cotton soil. Black soils are mainly found over the Deccan lava tract (Deccan Trap) including Maharashtra, Madhya Pradesh, Gujarat and Andhra Pradesh. These soils are formed by the weathering of igneous rocks and the cooling of lava after a volcanic eruption. Black soils are fine textured and clayey and have a characteristic dark colour, varying from dark brown to deep black. These soils are divided into:

Shallow black soils: Soils with a depth of 30cm or less.

Medium black soils: Soils with a depth of 30cm to 100cm.

Deep black soils: Soils with a depth of 100cm or more.

Soils: This is well known group of soils characterized by dark grey to black colour with high clay content. Soils are rich in montmorrillinite and beidilitic group of clay minerals, high CEC (40-60 me/100 g soil). These soils have high water holding capacity but have poor drainage and consequent water logging. These soils are classified in the order Entisols, Inceptisols and vertisols.

Nutrient status: Black colour of soil is due to the presence of titaniferous magnetite, organic compounds of iron and aluminium, accumulated humus and alluminium silicate. These soils are deficient in nitrogen, phosphorus, organic matter but rich in potassium, magnesium carbonates.

pH: 7 to 8.

Crops suitable under black soils: Cotton, millets, soybean, sorghum, pigeonpea, sugarcane, wheat, tobacco and citrus crops.

9. Desert soil: Desert soils occur under arid and semi –arid conditions

Soils: These soils are mostly sandy to loamy fine sand with brown to yellow brown colour. These soils have poor clay content and have less moisture. Soils consist mostly of sands.

Nutrient status: The desert sand is made up of quartz but feldspar and horneblende grains also occur with a fair proportions of calcareous grains. Salt content is too high. Nitrogen content is very low which is present in the form of nitrates. The presence of phosphate and nitrate make the desert soils fertile and productive under water supply.

pH: 7.2 to 9.2.

Crops suitable under desert soil : Gram, wheat, barley, oilseeds, maize etc.

10. Saline and alkali soils

Distribution: These soils occur in different parts of India like Uttar Pradesh, Haryana, Punjab, Maharastra, Tamilnadu, Gujarat, Rajastan and Andhra Pradesh.

Soil: Saline soils are considered better than the alkali soils as the soil is not dispersed and can be rectified with irrigation and Sulfur or Gypsum application while alkali soils can be reclaimed only by adding gypsum and in certain cases by scraping surface. These soils are classified under Aridisols, Entisols and Vertisols.

Nutrient status: Saline soils contain excess of natural soluble salts dominated by chlorides and sulphates which affects plant growth. Sodic or alkali soils contain high exchangeable sodium salts.The injurious salts are confined to the top layer of the soil.

Crops suitable under Saline and Alkali soils: Rice, maize, Bajra, barley and sorghum etc.

11. Peaty and other organic soils

Distribution: Peaty soils are found more in Kerala and marshy soils are found more in coastal tracks of Orissa, West Bengal and South-East coast of Tamilnadu, Almora in Uttarakhand, Central portion of North Bihar.

Soils: The soils of these places are generally blue due to the presence of ferrous irons and organic matter. These soils originated in humid regions as a result of an accumulation of large amount of organic matter in the soils. During monsoons the soils get submerged in water and the water receipts after the monsoon during which period rice is cultivated.

Nutrient status: Free aluminium and ferrous sulphate are present and poor in available phosphorus and are highly toxic to plant life as they contain ferrous and aluminium sulphates in considerable amounts.

pH: 3.5.

Crops suitable under Peaty and other organic soils : Rice.

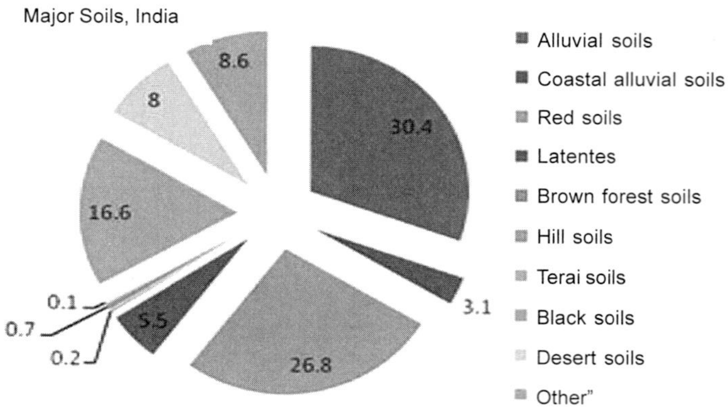

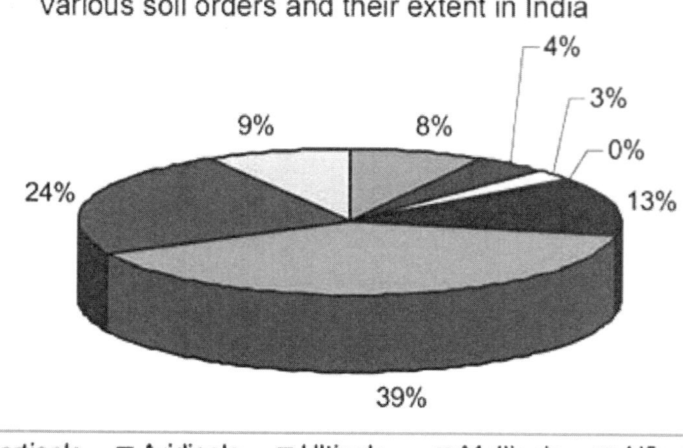

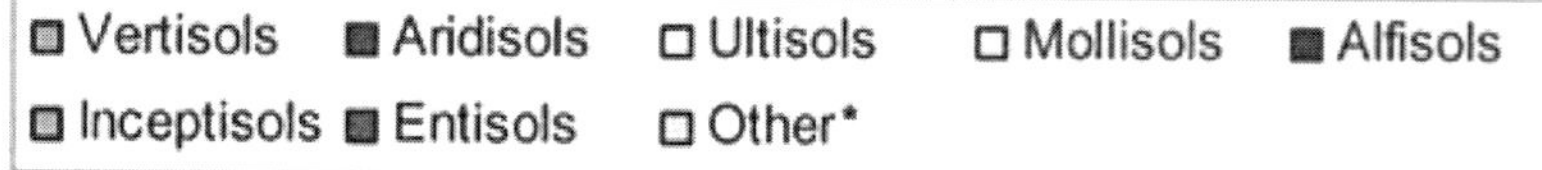

Chapter 4

Problematic Soils: Saline and Alkali Soils, Acidic Soils

The problem soils are those which, due to its land or soil characteristics, cannot be economically used for the cultivation of crops without adopting proper reclamation measures.

Classification of Problem Soils: The problem soils are divided into 3 groups.

- Highly Eroded Soils
- Saline and Alkali Soils
- Acid Soils

Highly Eroded Soils

These soils are called 'Usar' in U.P., 'Kallar or Thur' in Punjab and 'Luni' in Rajasthan. In India, area under saline alkali soils is 7.0 mha. When the chloride (Cl), sulphate (SO_4^{2-}), carbonate (CO_3^{2-}) and bicarbonate(HCO_3^-) salts of sodium (Na^+), calcium (Ca^{2+}) and magnesium (Mg^{2+}) are increased in the soil, the soil become saline and alkaline. On the basis of amount of soluble salts, average quantity of exchangeable Na and pH, they are classified into 3 groups

- Saline soils
- Alkali soils
- Saline – alkali soils

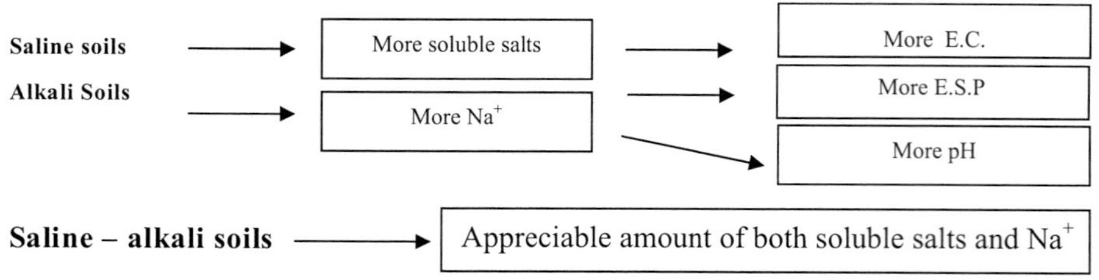

A. Saline and Akali Soils

Characteristics

pH: < 8.5

EC (Electrical conductivity): > 4.0 m.mhos/ cm or decimens/metre) at 25°C.

ESP (Exchangeable sodium %): >15

Detrimental effect of soil salinity and soil alkalinity

- Less water permeability
- Low aeration and low microbial activities
- Unavailability of Phosphorus, calcium and nitrogen
- Hinderance in water absorption
- Increase in osmotic pressure adversely effects the plant growth
- Soil is dispersed and become compact.

Geographical Distribution

- **Soils of Arid alluvial belts:** Western part of Punjab, Haryana, Rajasthan, Rann of Kutch (Gujarat).
- **Soils of Semi arid regions:** Punjab, Haryana, and West Uttar Pradesh.
- **Arid and Semi arid region:** Deccan Pleateau which is in the area between Narmada and Godawari.
- **Along sea coasts and eastern parts of India**, where heavy rains reduce the problem.
- **Western delta of Ganga and Cauvery river.**

Causes for the formation of saline-alkali soils

I. Geological factors

- Weathering of rocks and minerals

- Tidal waves
- Use of poor quality of irrigation water

II. Man-made problems

- Use of excessive irrigation
- Poor drainage system
- Use of fertilizers

III. Climatic factors

If the Evaporation is more than rainfall, then there is development of saline-alkali soils.

B. Saline or White Alkali Soils

The saline soils contains toxic concentration of soluble salts in the root zone. Soluble salts consists of chlorides and sulphates of sodium, calcium, magnesium. Because of the white encrustation formed due to salts, the saline soils are also called white alkali soils

Types of saline soils

- Having Ca and Mg.
- Having Na which damages colloidal complex.

Characteristics

pH: < 8.5

EC (Electrical conductivity) : >4.0 m.mhos/cm or decimens/metre) at 25°C.

ESP (exchangeable sodium %): <15

Chemistry of soil solution: Dominated by sulphate and chloride salts of Na, Ca and Mg.

Formation of Saline Soils

i. Common in arid and semi-arid region where annual rainfall is less than 55 cm.

ii. Under low rainfall condition the soluble salts formed by weathering are not completely leached out of the soil and get concentrated in the surface layers.

iii. In areas having salt layer at lower depths in the profile, seasonal irrigation may favour the upward movement of salts.

iv. Salinity is also caused if the soils are irrigated with saline water.

v. In coastal areas, the ingress of sea water induces salinity in the soil.

vi. In dry conditions, the salts move up with water and brought to the surface and deposited there when water evaporates.

Favourable condition of saline soil formation

- Low rainfall
- High water table with high salt concentration
- High temperature

Effect on crops and soil

- The soluble salts accumulate and increase the osmotic pressure of the soil solution and plant is unable to absorb sufficient water for meeting their physiological requirement. While on the other side, there is diffusion of water from the plants into the soil through the roots cells which are in contact of the concentrated soil solution. The protoplasm begins to shrink and there is gradual wilting of plants.
- Development of thicker layer of surface wax imparts bluish green tinge on leaves.
- Due to high EC, germination percent of seeds is reduced.

Amelioration

Main principle is removal of excess salt to a desired level in the root zone. Amelioration consists of application of sufficient quantities of good quality water to the soil, to dissolve the salts and leach them out through drainage channels. Soluble salts leach down in about 3-4 days of flooding.

Method: The affected area is to be made into smaller plots and each plot should be bunded to hold irrigation water. Separate irrigation and drainage channels are to be provided for each plot. Then, the plots are to be flooded with good quality water upto 15 - 20 cm and puddled. Thus, soluble salts will be dissolved in the water. Thereafter, excess water with dissolved salts is removed through the drainage channels. This process of flooding and drainage is repeated 5- 6 times till all the soluble salts are leached from the soil to a safer limit.

Crops suitable for cultivation in saline soils

Rice, Maize, Redgram, Greengram, Sunflower, Linseed, Sesame, Bajra, Sorghum, Barley, Sugarbeet, Cotton, Sugarcane, Mustard, Tomato, Cabbage, Cauliflower, Cucumber, Pumpkin, Bitterguard. Beetroot, Guava, Asparagus, Banana, Spinach, Coconut, Grape, Datepalm, Pomegranate.

C. Alkali Soils

These soils are also called Sodic or Black Alkali or Non-saline-alkali soils. In these soils, colloidal complex is saturated with exchangeable sodium and absence of appreciable quantities of soluble salts. These soils are black in colour which is due to sodium content which causes dispersion of organic matter. According to Kelley, black alkali always contain a large amount of absorbed sodium whereas white alkali may or may not contain this material.

Formation of Sodic soils

i. Common in areas where annual rainfall is 55–90 cm and low lying areas with inadequate drainage.

ii. When the soluble salts accumulating in soils include some sodium, it replaces calcium in the clay complex and sodium clay is formed.

iii. The excessive irrigation of uplands containing Na salts results in the accumulation of salts in the valleys.

iv. In arid and semi arid areas salt formed during weathering are not fully leached.

v. In coastal areas if the soil contains carbonates, the ingression of sea water leads to the formation of alkali soils due to formation of sodium carbonates.

Evolution of Alkali soils is divided into three distinct stages

a. Salination

b. Saline – alkaline soils

c. Alkalization i.e.re-desalination and intense alkali soil formation.

Classification of Alkali Soils: 3 Classes

a. **Class I :** Soils having alkali earth carbonates.

b. **Class II :** Soils having pH > 7.5 but no alkali earth carbonates.

c. **Class III :** Soils having pH < 7.5 but no alkali earth carbonates (Degraded alkali).

Characteristics

pH: > 8.5

EC (Electrical conductivity) : <4.0 m.mhos/ cm or decimens/metre) at 25°C.

ESP (Exchangeable sodium %): >15

Chemistry of soil solution: Dominated by carbonate and bicarbonate ions and high exchangeable sodium

Amelioration

Principle: Replacement of exchangeable Na^+ and Ca^{2+} and leaching of Na^+ salt out of the root zone. Soils get dispersed and become compact. In these soils, exchangeable Na–clay make impervious to water. Thus, Na^+ must be replaced by Ca^{2+} and then leached downwards.

Amendments

1. **Application of gypsum** ($CaSO_{4.}2H_2O$)

Clay Complex

$$\begin{matrix} Na^+ \\ \boxed{\text{Clay Complex}} \\ Na^+ \end{matrix} + CaSO_4 \longrightarrow \begin{matrix} Ca^2 \\ \boxed{\text{Clay}} \\ Ca^2 \end{matrix} + Na_2SO_4 \downarrow$$

Alkali Soil Gypsum Normal Soil

$$Na_2SO_4 + CaSO_4 \longrightarrow Ca\ CO_3 + Na_2SO_4 \downarrow$$

Sodium sulfate goes into the soil solution and leaches down.

Gypsum application: Gypsum is applied in the soil surface@ 12-15 t /ha and mixed by harrowing 2 to 4 weeks before sowing.

2. Application of sulphur: Sulphur is oxidized to sulfuric acid by a group of organisms called sulphur bacteria. Sulphur is applied @ 2.5-4 tons/ha

$$2S + 3O_2 + 2H_2O \xrightarrow{\text{Sulfonification}} 2\,H_2SO_4$$

$$H_2SO_4 + Na_2CO_3 \xrightarrow{\text{Microbial}} Na_2SO_4\downarrow + H_2O + CO_2$$

$$H_2SO_4 + CaCO_3 \rightarrow CaSO_4\downarrow + H_2O + CO_2$$

Ground sulfur is incorporated into soil at the rate of 2.5 – 4 tons/ha several weeks before planting of the crop.

3. **Use of sulfuric acid:** $Na_2\,CO_3 + H_2SO_4 \rightarrow Na_2SO_4\downarrow + H_2O + CO_2$

4. **Use of iron sulphate:** $FeSO_4 + H_2O \rightarrow H_2SO_4 + FeO$

$$Na_2\,CO_3 + H_2SO_4 \rightarrow Na_2SO_4\downarrow + H_2O + CO_2$$

5. **Use of iron pyrite:**

$$2FeS_2 + 7O_2 + 2H_2O \rightarrow 2FeSO_4 + 2H_2SO_4.$$

$$CaCO_3 + H_2SO_4 \rightarrow CaSO_4 + H_2O + CO_2$$

6. **Use of lime sulphur:**

$$CaSO_4 + 4\,H_2O + SO_2 \rightarrow CaSO_4 + 4\,H_2SO_4$$

Gypsum Requirement (GR) : The amount of gypsum required to lower the ESP of a sodic soil to a desired level is called Gypsum Requirement.

$$GR = (C_1 - C_2)\ X\ 2\ m\ eq\ of\ Ca^{2+}/100\ g\ soil$$

C_1 = Concentration (m eq l^{-1}) of Ca^{2+} in gypsum solution

C_2 = Concentration (m eq l^{-1}) of Ca^{2+} and Mg^{2+} in filtrate

Management Factors

1. Increasing Nitrogen doses
2. Green manuring: Rice-Dhaincha , Dhaincha-Rice-Berseem
3. Light irrigation with frequent interval
4. Use of salt free irrigation water
5. Application of lime and organic matter
    a. When soil is deficient in lime, its application helps in reclamations.
    b. Application of organic matter increases biological activity and carbon dioxide production in the soil. It assists to dissolving the calcium carbonate and leaching down the bicarbonate solution to lower layers. At the same time, sodium carbonate is also converted to bicarbonate, which result in lowering down pH.

6. Use of Alkali resistant crops : Barley, sugarbeet, cotton, sugarcane, mustard, rice, barley, oat, wheat, berseem, tomato, potato, carrot, cauliflower, brinjal, dates, ber, orange, citrus.

Effect on Crops and Soil

- Poor condition of soil due to dispersion of soil particles due to high exchangeable 'Na'. There is low permeability to water and air. Soils tends to be sticky when wet and becomes hard on drying.
- High exchangeable sodium decreases the availability of calcium and magnesium to plants. It also effects the solubility of zinc (Zn).
- Toxicity due to excess hydroxyl and carbonate ions.
- Growth of plant gets affected mainly due to nutritional imbalance.
- Root system is retarded and there is delay in flowering.
- There is burning of leaves in annuals and woody plants due to excess of chloride and sodium.
- Bronzing of leaves in citrus plants.
- Application of molasses.
- Drainage channels must be arranged around the field.
- Growing the green manure crops and incorporate in the field.

Table: Classification on the basis of soil salinity tolerance

Saline sensitive crops	Less salinity tolerant	High salinity tolerant
Electrical conductivity (EC) < 4.0 m mhos/cm Beans, radish,carrot, pear, apple, orange, almond, citrus	Electrical conductivity (EC) 4-10 m mhos/cm Wheat, oat, rice, maize, linseed sunflower, soybean, tomato, cabbage, cauliflower, potato,chilli onion, cucumber, bottlegourd grape, mango, pomegranate, sudangrass	Electrical conductivity (EC) >10 m mhos/cm Barley, sarson, cotton, beet, palak, date

Table: Classification on the basis of soil alkalinity tolerance

Sensitive crops	Low tolerant crops	Tolerant crops
ESP<15% Beans, Maize, carrot, orange, peach	ESP=15-40% Carrot, oat, onion, radish rai, jowar, wheat, pea spinach	ESP> 40% Oat, beet, cotton, Rhodes grass

Humid alkali : It is very less in area while problem is more acute than dry regions. The alkali soils of humid regions are confined in the main to swampy areas or spots where seepage may have brought soluble materials to the surface. Through evaporation, the soluble salts accumulate in the surface soil and persist inspite of leaching.

Reclamation : 3 methods are suggested

- Drainage.
- Application of potash salts and phosphorus.
- Addition of trash organic matter.

Table: Difference between saline and alkaline soils

S.No.	Saline soils	Alkaline soils
1.	These soils are also known as white alkali soils.	These soils are also called Sodic or Black Alkali or Non-saline-alkali soils.
2.	The saline soils contains toxic concentration of soluble salts in the root zone. Soluble salts consists of chlorides and sulphates of sodium, calcium, magnesium.	The colloidal complex is saturated with exchangeable sodium and absence of appreciable quantities of soluble salt in Alkaline soils
3.	These soils are slightly alkaline and pH is less than 8.5.	These soils are slightly alkaline and pH is more than 8.5.
4.	Electrical conductivity of their saturation extract is more than 4.0 m.mhos/ cm or decimens/metre) at 25°C.	Electrical conductivity of their saturation extract is less than 4.0 m.mhos/ cm or decimens/metre) at 25°C.
5.	The soil contains more than 0.2 percent harmful soluble salts.	The soil contains less than 0.2 percent harmful soluble salts.
6.	The soil contains < 15% exchangeable sodium percentage (ESP).	The soil contains >15% exchangeable sodium percentage (ESP).
7.	Flocculated soils therefore aeration and permeability is normal.	Dispersed and compact soil, aeration and permeability is low.
8.	The physical conditions remain favourable for cultivation.	The physical conditions is very poor and is unfit for cultivation.
9.	These soils can be reclaimed by mechanical method upto some extent.	Use of amendments is must for reclaiming the soils.
10.	Organic matter is present in soil.	Organic matter is absent in soil.

Usar Soils

Usar soils are present in Punjab, Haryana and Uttar Pradesh.

Reclamation: Usar soils can be reclaimed by following five methods:

- Soil management
- Crop management

- Chemical methods
- Mechanical methods
- Biological methods

1. Soil management

- Irrigation :
  i. Good quality of irrigation water. Avoid use of salty water.
  ii. Irrigation should be done timely.
- Drainage : There should be good drainage system in order to remove the salts from the root zone.
- Decrease evaporation rate from soil surface by the use of mulches, crop residues etc.
- Use of organic manures : eg, FYM, Compost, green manure improve the soil fertility.
- Use of Acidic fertilizers eg. Ammonium sulfate or super phosphate.
- Light irrigation at frequent intervals.

East-West bunds are made and crops are sown on the northern side.

2. Crop management

i. **Selection of suitable crops**

Salt tolerant crops

Low salt tolerant	-	Pulses, radish, beans, sesame
Medium salt tolerant	-	Wheat, maize, cotton, sorghum, mustard
High salt tolerant	-	Rice, sugarcane, oats
Alkali tolerant crops		
Alkali sensitive	-	Maize, bean, carrot, orange, peach
Low Alkali tolerant	-	Wheat, jowar, oat, pea, onion, radish, carrot
Alkali tolerant crop	-	Cotton, oat, beetroot, Rhodes grass

ii. **By growing trees**: Acacia by litter fall and taproot system of trees, there is breaking of hard pans of soils.

3. Chemical method : As listed in Sodic Soils.

4. Mechanical methods

i). Flooding and leaching down the salts. While leaching, there should not be hard pan in soil and there should be low water table of at least 3 metre deep.

ii). Scraping of salts from the surface of soil.

iii). Underground Drainage: By laying artificial drains under the soil and salts come to these drains by seepage and water is drained out.

iv). Trenching : Trenches are made for draining salty water.

5. Biological method

Use of green algae after flooding and then application of FYM. These practices improves the soil structure, and there is no more compaction.

Acid Soils

Acid soils are those soils in which pH less than 6.5 and which respond to liming.

Distribution of Acid Soils

Out of 157 million hectares of cultivable land in India, 49 million hectare of land are acidic. Acid soils are present in almost all major groups of soil except black soil (vertisols). These soils are found in North-east region, eastern and peninsular regions, coastal plains, coastal alluvium, peaty and marshy soil, Himalayan regions.

Nature of Soil Acidity

There are two types of soil acidity

1. **Active Acidity:** Active acidity may be defined as the acidity developed due to hydrogen (H^+) and aluminium $(Al)^+$ ions concentration of the soil solution. The magnitude of this acidity is limited.
2. **Reserve/Potential/Exchange Acidity:** Exchange acidity may be defined as the acidity developed due to adsorbed hydrogen (H^+) and aluminium (Al^{3+}) ions on the soil colloids. These ions are not so free to move about as are those in the soil solution, but they are in equilibrium with them. The magnitude of this exchange acidity is very high.

Total acidity = Active acidity + Exchange acidity. Therefore, total soil acidity depends on the active and the exchange acidity of the soil.

Classification of Soil pH Ranges

The United States Department of Agriculture, formerly Soil Conservation Service classifies soil pH ranges as follows:

Denomination	pH range
Ultra acid	<3.5
Extremely acid	3.5 - 4.4
Very strongly acid	4.5 - 5.0
Strongly acid	5.1 - 5.5
Moderately acid	5.6 -6.0
Slightly acid	6.1 -6.5
Neutral	6.6 - 7.3
Slightly alkaline	7.4 - 7.8
Moderately alkaline	7.9 - 8.4
Strongly alkaline	8.5 -9.0
Very strongly alkaline	>9.0

Factors for formation of Acidic Soil

1. **Leaching of Ca and other bases due to heavy rainfall:** Water passing through the soil leaches basic cations such as calcium (Ca^{2+}), magnesium (Mg^{2+}), and potassium (K^+) into drainage water. These basic cations are replaced by acidic cations such as aluminum (Al^{3+}) and hydrogen (H^+). For this reason, soils formed under high rainfall conditions are more acid than those formed under arid conditions. In areas where rainfall is more than 1,000 mm, acidic soils are more common.

$$CO_2 + H_2O = \underset{\text{(Carbonic acid)}}{H_2CO_3}$$

$$H_2CO_3 + CaCO_3 = \underset{\text{(Leachable)}}{Ca\ (HCO_3)_2}$$

2. **Formation of soil with acidic parent :** Some soils have developed from parent material which is acidic in nature. Rocks like granite and Rhyolite produce silisic acids mainly ortho silisic acid and trisilisic acids.

3. **Removal of basis by crops :** Plants take up basic cations such as K^+, Ca^{++}, and Mg^{++}. When these basic cations are removed from the soil, they are replaced with H^+ in order to maintain electrical neutrality. Sugarbeet and potato absorbs salts from soil and reduces basis in the soil.

4. **Use of Acid forming fertilizer:** Both chemical and organic fertilizers may eventually make the soil more acid. Hydrogen is added in the form of ammonia-based fertilizers (NH_4^+) , urea-based fertilizers $CO(NH_2)_2$ and as proteins (amino acids) in organic fertilizers. Transformations of these sources of N into nitrate (NO_3^-) releases H^+ to create soil acidity. Therefore, use of fertilizers containing ammonium or even adding large quantities of organic matter to a soil will increase the soil acidity and lower the pH.

$$NH_4^+ + 2O_2 \xrightarrow{\text{Bacteria}} NO_3^- + 2H^+ + H_2O$$

5. **Microbial activities:** Micro-organisms decompose organic matter in the soil and organic acids are continuously being formed, making the soil acidic.

$$CO_2 + H_2O \rightarrow H_2CO_3$$

$$H_2CO_3 + CaCO_3 \rightarrow Ca\,(HCO_3)_2$$

6. **Farm practices :** Leaching of salts in fallow land causes acidity.

Conditions Favourable for Soil Acidity: There are three conditions of soil acidity.

a. Strongly acid soils : In these soils, pH is less than 5.0 and aluminium become soluble and is either tightly bound by organic matter or is present in the form of Aluminium or aluminium hydroxyl cations.These exchangeable ions are absorbed by the negatively charged soil colloids. The adsorbed aluminium is in equilibrium with aluminium ions in the soil solution, and the aluminium ions contribute to soil acidity through their tendency to hydrolyze.

$$\boxed{\text{Colloid}}\ Al^{3+} \rightleftharpoons Al^{3+}$$

Adsorbed aluminium in soil colloid	Aluminium ion in soil solution

In soil solution, aluminium ion produce hydrogen ion by following hydrolysis reaction.

$$Al^{3+} + H_2O \rightarrow Al\,(OH)^{2+} + H^+$$

b. Moderately acid soils : pH ranges from 5.0 - 6.5. These soils have higher percentage of base saturation (Ca^{2+},Mg^{2+} etc.) than strongly acid soils. In the soil solution, aluminium hydroxyl ions produce hydrogen ions by the hydrolysis reaction.

$$Al\,(OH)^{2+} + H_2O \rightarrow Al\,(OH)^+_2 + H^+$$

$$Al\,(OH)_2^+ + H_2O \rightarrow Al\,(OH)_3 + H^+$$

c. Acid sulphate soils

In case of these soils (pH < 3.5), the acidity is due to dissolved or free acidic substances, such as sulphuric acid, ferric and aluminium sulphate. The sulphuric acid is produced by oxidation of sulphur and sulphide. These soils occur in some of low lying areas of Kerala.

Characteristics of Acid soils : These are classified into three groups.

a. Physical

- Light texture
- Poor organic matter
- High permeability
- Poor Cation Exchange capacity
- Low water holding capacity
- Mainly kaolinite and sometimes illite clay minerals.

b. Chemical

- Active and potential soil acidity
- Base unsaturated soil means there are more anions than cations.

Table : Effect of pH on Nutrient availability

pH	Available Nutrient	Remarks
Neutral pH	P,B	Boron is available at pH 5.0-7.0
Low pH	Al, Fe, Mn, Zn, Cu, Co	All trace elements except B and Mo
High pH	Mn, N,K,Ca,Mg,S	All major nutrients except 'P'

c. Biological

- Fungi are higher than bacteria. There is no effect of pH on fungi.
- Fungi causes diseases.
- Decomposition rate and rate of mineralization and utilization is reduced when acidity increases.

Problems in Very Acid soils

- Aluminium toxicity to plant roots.
- Manganese toxicity to plant roots.
- Calcium and magnesium defiency.
- Molybdenum deficiency in legumes.
- Phosphorus tied up by Fe and Al.
- Poor bacterial growth.
- Reduced nitrogen transformations.

Effect of soil acidity on Plants

Plants are effected by acidity both directly and indirectly.

Direct effects

1. It has toxic effect on root tissues and adversely affects the permeability.
2. It disturbs the balance between basic and acidic constituents of the plants and affects growth of plants.
3. It affects enzymatic changes which are particularly sensitive to pH changes.
4. If affects the beneficial activity of soil micro-organisms.

Indirect effects

1. High solubility and availability of elements like Al, Mn and Fe in toxic amounts due to high soil acidity
2. Due to soil acidity, Ca and K may be deficient.
3. It affects the availability of phosphorus, copper and zinc.
4. Plant diseases occur more in acidic soils.
5. Disturbance in balance between acidic constituent through the roots.

Management of Soil acidity: It can be done by two ways.

- By growing acid tolerant crops.
- By application of soil amendments.

Liming : Adding to the soil any compound containing Ca alone or both Ca and Mg that is capable of reducing soil acidity is called liming. Lime reduces soil acidity (increases pH) by changing some of the hydrogen ions into water and carbon dioxide (CO_2). A Ca^{++} ion from the lime replaces two H^+ ions on the cation exchange complex. The carbonate (CO_3^-) reacts with water to form bicarbonate (HCO_3^-). These react with H^+ to form H_2O and CO_2. The pH increases because the H^+ concentration has been reduced.

Remember, the reverse of the above process can also occur. An acid soil can become more acid as basic cations such as Ca^{2+}, Mg^{2+}, and K^+ are removed, usually by crop uptake or leaching, and replaced by H^+.

Liming material: The most common liming materials are $CaCO_3$, CaO (burnt lime), $Ca(OH)_2$ (slaked lime), Dolomite (rich in Mg), Coral shell, Chalk ($CaCO_3$), Blast furnace slag $CaSiO_3$ and Ca_2SiO_4.

$$\text{Soil}\,(H^+, H^+) + CaCO_3\ (\text{Lime}) \rightarrow \text{Soil}\,(Ca^{++}) + H_2O\ (\text{Water}) + CO_2\uparrow\ (\text{Carbon Dioxide})$$

Amount of lime to supply depends upon

- pH or intensity of the soil acidity
- Texture of the soil
- Lime requirement of crops
- Purity of material
- Chemical composition of material
- Fineness of particle size

Method of applying lime: Lime should be mixed in upper 10 cm layer of soil, before ploughing or on ploughed land well before cropping and should be thoroughly mixed with the soil by disking. Efficient way to use lime is to apply it in small quantities every year or once in two years. If the quantity is not more than 5 tons/ ha it can be applied in one dose. If the dose is more one half is applied before ploughing and the remaining half applied and mixed in after ploughing.

Lime Requirement: It is generally a measure of base (lime) required to neutralize that fraction of the total acidity that must be neutralized to attain a desired soil pH that is favourable for crop growth.

In order to raise the pH by one unit i.e., from 5.0 to 6.0 in one hectare about 1,500 kg., of $CaCO_3$ would be required. Lime requirement depends on:

1. Texture of the soil: Clay loam soils require more lime than sandy soil.
2. Purity of the liming material.
3. Degree of fineness: Finer the material is ground, more rapidly it goes in solution and is more effective. Material passing through 60 mesh sieve is considered to be standard and quite effective.
4. Chemical composition: The neutralizing value of $CaCO_3$ is taken as 100.

Soils having higher organic matter require more lime.

Effect of liming on Soil and Plant

- Liming reduces soil acidity and increase calcium content in soil solution
- Improves physical condition of soil
- Beneficial soil bacterial are encouraged
- Lime makes phosphorus more available.

Effect of over liming

- Non availability of Boron, Fe, Mg, Cu and Zn
- Growth of Thielavia basicola fungus - Diseases is more.

Other measures

1. By growing acidic crops suitable for particular pH

   For medium acidic soil (pH upto 5.6): Jowar, maize, wheat, sweet potato, tobacco, arhar, velvet bean, turnip.

   For strongly acidic soils (pH upto 5.1): potato, rice, oats, rye, cowpea, linseed, grasses, tea and coffee.

2. By growing acid tolerant crops like alfalfa, bean, grasses, beet, cauliflower, cabbage, clover, maize, rice, potatoes, barley, soybean, sunflower.

Common Terms

- Exchangeable Sodium Percent (ESP) or Soluble Sodium Percentage (SSP)

$$ESP = \frac{\text{Exchangeable Sodium (in milli equivalent / 100 g soil)}}{\text{Total Cation Exchange Capacity (in milli-equivalent/100 g soil)}} \times 100$$

- Sodium Absorption Ration

$$SAR = \frac{[Na^+]}{\frac{\sqrt{Ca^{2+} + Mg^{2+}}}{2}}$$

When Na^{2+}, Ca^{2+}, Mg^{2+} are concentration in c mol/kg in soil solution.

Chapter 5

Nutrient Management

Plant Nutrition

The term "fertility" refers to the inherent capacity of a soil to supply nutrients to plants in adequate amounts and in suitable proportions. Nutrition in plants may thus be defined as a process of synthesis of food, its breakdown and utilisation for various functions in the body. There are two types of nutrition:

1. Autotrophic nutrition: It is a type of nutrition in which the living organisms manufacture their own organic food from simple inorganic raw materials. The green plants exhibit autotrophic mode of nutrition and hence called the autotrophs.

2. Heterotrophic nutrition: Certain non green organisms like fungi and many bacteria fail to synthesize their own organic nutrients from inorganic substances. These organisms are thus dependent on some other external sources for their nutrition. Such plants are called heterotrophic plants and the mode of nutrition is called heterotrophic nutrition.

Essential Elements

Plant absorb more than 90 elements from soil, water and air. Out of these elements, 17 are essential for growth of most plants. These are called Essential elements. The term essential mineral element (or mineral nutrient) was proposed by Arnon and Stout (1939). Arnon in 1954 laid down the criteria of essentiality. These criteria are:

1. Plant must be unable to grow normally or complete its life cycle in the absence of the element.
2. The element is specific and cannot be replaced by another element.
3. The element play a direct role in metabolism.

Recent investigations show that the point (2) cannot be accepted absolutely as Mo, may be substituted by Vanadium (V); similarly Chlorine (Cl) by Bromine (Br); Potassium (K) by Rubidium (Rb); calcium (Ca) by Strontium (Sr).

Uptake of Mineral Elements

Plants absorb a large number of minerals from soil. The uptake of mineral ions by the roots may be passive or active.

a. **Passive absorption :** It is the initial and rapid phase and ions are absorbed into the "outer space" of the cells, the apoplast. It does not require use of any metabolic energy.

b. **Active absorption :** It is the second phase of ion uptake. The ions are taken in slowly into the 'inner space' the symplast of cells. It needs the expenditure of metabolic energy.

The movement of ions is called flux. When the ions move into the cells, it is called influx and the outward movement of ions is called efflux. The minerals ions absorbed by the root system are translocated through the xylem vessels to other parts of the plant.

Classification of Nutrients

There are different basis of classification of essential nutrients.

A. Quantity of nutrient required

B. Mobility of nutrient in soil

C. Mobility of nutrient within plant

D. Functions in plant.

A. Classification on the basis of quantity of nutrient required

i. **Basic nutrients:** These constitute 96% of total dry matter of plant. These are carbon, hydrogen and oxygen. Among these, carbon and oxygen constitute 45% each and hydrogen is 6%.

ii. **Macronutrients:** The nutrients which are required in large amounts i.e. >1 ppm for their growth and survival are called macronutrient. These are nine in numbers *eg.* nitrogen, phosphorus, potassium, calcium, magnesium, sulphur, carbon, hydrogen and oxygen. These macronutrients, again are divided into two groups.

Primary nutrients : They are nitrogen (N), phosphorus (P) and potassium (K).

Secondary nutrients: They are Calcium (Ca), magnesium (Mg) and Sulphur (S), which are present in soil in adequate amount, so fertilization is not always needed. During liming of acidic soils, large amount of calcium and magnesium are added and sulphur is available from slow decomposition of organic matter.

iii. **Micronutrients:** They are also called minor/trace/oligo/spurne elements- The nutrients which are required in very small (micro) quantities (<1 ppm) are called micronutrient. The micronutrients are boron (B), copper (Cu), iron (Fe), chloride (Cl), manganese (Mn), molybdenum (Mo) and Zinc (Zn). Recently, Nickel (Ni) has also been added as essential element. Efficiency of micronutrients is very high as their small amount produce good results, while their deficiency caused harmful effect.

Beneficial elements /Potential micronutrients: Those nutrients which are beneficial to plants but not essential. They are cobalt (Co), nickel (Ni), selenium (Se), aluminium (Al), Rubidium (Ru), strontium (Sr), chromium (Cr), sodium (Na), arsenium (As), vanadium (Va) and silicon (Si). These elements stimulate the plant growth even at low concentrations.

Ultra nutrients: Plant require in very small quantity *i.e.* 1 ppb (parts per billions). Nicholas (1963) put forward this concept giving example of Mo and Co but it is faulty classification and is not accepted today because the quantity of nutrients absorbed depend upon the plant type.

B. Classification on the basis of mobility of nutrient in the soil

Mobile nutrients: The nutrients are highly soluble and these are not adsorbed on clay complexes. Example: NO_3^-, SO_4^{2-}, BO_3^{3-}, Cl^- and Mn^{+2}

Less mobile nutrients: They are soluble, but they are adsorbed on clay complex, so their mobility is reduced.

Example: NH_4^+, K^+, Ca^+, Mg^{2+}, Cu^{2+}

Immobile nutrients: Nutrient ions are highly reactive and get fixed in the soil.

Example: $H_2PO_4^-$,HPO_4^{2-}, Zn^{2+}

C. Classification on the basis of mobility within plant

Highly mobile:	N, P and K.
Moderately mobile:	Zn
Less mobile:	S, Fe, Mn, Cl, Mo and Cu
Immobile:	Ca and B

D. Classification on the basis of functions in the plant

- Elements that provide basic structure to plant.

  Example: Carbon, Hydrogen and Oxygen.
- Elements useful in energy storage, transfer and bonding: These are accessory structural elements which are more active and vital for living tissues. Example: N, S and P.
- Elements necessary for charge balance. Example: K, Ca and Mg.
- Elements involved in enzyme activation and electron transfer. Example: Fe, Mn, Zn, Cu, B, Mo and Cl.

Beneficial nutrients: These are not included in essential nutrients, but their application increases the yield up to some extent.
Example: Sodium, Silicon, and Vanadium.

Table: Form of elements absorbed by the plants

Elements	Ionic form	Non Ionic form	Source	Name of the discoverer and Year
1. Carbon	$C0_3^{-2}$, HCO_3^-	CO_2	Air and water	Sachs, 1882
2. Hydrogen	H^+, OH^-	H_2O	Air and water	Sachs, 1882
3. Oxygen	OH^-	O_2	Air and water	De Saussure, 1804
4. Nitrogen	NO_3^- , NH_4^+	$CO(NH_2)_2$	Parts of N from air but mostly from soil	Ruther Ford, 1872
5. Phophorus	HPO_4^{-2} $H_2PO_4^-$	Nuclei acid, Phytin		Posternack, 1903
6. Potassium	K^+	-	Soil	Schimper, 1890
7. Calcium	Ca^{+2}		Soil	Solm Harstmax, 1856
8. Magnesium	Mg^{+2}			Willstutter, 1906
9. Sulphur	SO_4^{-2}	SO_2 from air	Soil and air	Peterson,1911
10. Iron	Fe^{2+}, Fe^{3+}	$FeSO_4$ with EDTA	Soil Soil	Gris,1844
11.Copper	Cu^{+2}, Cu^+	$CuSO_4$ with EDTA	Soil	Brenchley, 1914
12. Zinc	Zn^{+2}	$ZnSO_4$ with EDTA		Mazla, 1915
13. Manganese	Mn^{+2}	$MnSO_4$ with EDTA	Soil	G. Betrand, 1897
14. Molybdenum	$HMoO_4^-$ MoO_4^{-2}	--	--	Arnon & Stout, 1939
15. Boron	$H_2BO_3^{-1}$ $B_4O_7^{-2}$, BO_3^{3-} HBO_3^{2-}	--	--	Agulhon,1910
16. Chlorine	Cl^-	--	--	TC Broyler, 1954
17. Sodium	Na^+	--	--	
18. Cobalt	Co^{++}	--	--	
19. Silicon	$Si(OH)_4$			

Nutrients mobility

In soil	
Mobile nutrients (highly soluble, not adsorbed on clay)	NO_3^-, Cl^-, Mn^{+2}, SO_4^-, BO_3
Less mobile nutrients (soluble/also adsorbed on clay)	NH_4^+, K^+, Ca^{2+}, Mg^{2+}, Cu^{2+}
Immobile (highly reactive/adsorbed in soils)	HPO_4^{2-}, $H_2PO_4^{2-}$, Zn^{2+}
In Plants	
Highly mobile	N,P,K,Mg,Mo
Moderately mobile	Zn
Less mobile	S,Fe,Cu,Mn
Immobile	Ca,B

Plant deficiency symptoms

- If symptoms occur in the old leaves, then it is likely the plant is deficient in a mobile element
- If symptoms occur in the young leaves, then it is likely the plant is deficient in an immobile element
- However, the deficiency may be due to a variable mobile element (where the symptoms may occur in either the young or old leaves)

The deficiency symptoms can be distinguished based on:

1. The region of occurrence
2. Presence or absence of dead spots
3. Chlorosis of entire leaf or interveinal chlorosis.

Table1: Deficiency Symptoms on the basis of region of occurrence, presence or absence of dead spots or chlorisis of entire leaf or interveinal chlorosis

S. No. Region of Occurrence	Deficient nutrient	Deficiency symptoms
1. **Old leaves**	N,P,K,Mg,Mo	
a. Without dead spots	N,P,Mg	
i. Yellow veins	N	**Nitrogen (N):** Reduced growth, uniform yellowing of leaves including the veins; leaf erect and stiff reduced lateral breaks; symptoms appear first on older growth
ii. Green veins	Mg	**Magnesium (Mg):** Yellowing in between the veins and veins remain green. Leaf is erect; reduction in growth; yellowish, bronze, or reddish color of older leaves, while veins remains green; leaf margins may curl downward or upward with a puckering effect.

iii. Leaves dark green,	P	**Phosphorus (P) :** Reduced growth : Leaves purple or red colordark green; purple or red color in older leaves, especially on the underside of the leaf along the veins ; leaf shape may be distorted; thin stems; limited root growth
b. With dead spots	K,Mo	**Potassium (K):** Yellowing start from tips or margins of leaves extending to the centre of the leaf base. The yellow part become nacrotic very soon. Reduced growth; shortened internodes; margins of older leaves become chlorotic and burn; necrotic (dead) spots on older leaves; reduction of lateral breaks and tendency to wilt readily; poorly developed root systems; weak stalks. **Molybdenum (Mo):** Translucent spots of irregular shape in between the veins. The spots are light green, yellow or brown in colour; twisted leaves (whiptail); marginal scorching and rolling or cupping of leaves.
2. **New leaves**	S, Fe, Mn, Cu	
i. Veins remaining green	Fe,Mg	**Iron (Fe):** Interveinal chlorosis; primary green vein remaining green and other portion of leaf turn yellow tending toward whiteness. Under severe condition it becomes white. **Manganese (Mn):** :Interveinal chlorosis; Both primary and smaller veins remain green; Interveinal portion white and not tending towards whiteness (grey or tan spots) usually develop in chlorotic areas; dead spots may drop out of the leaf; poor bloom size and color.
ii. Veins remaining yellow	S,Cu	**Sulfur (S):** Rarely deficient; general Yellow yellowing of the young leaves thereafter entire plant; veins lighter in color than adjoining interveinal area; roots and stems are small, hard and woody. **Copper (Cu):** Leaf is yellow tending toward whiteness. Chlorosis of veins occur and leaf looses its lusture. Leaf unable to retain turgidity and hence wilting occurs.
3. **Old and new leaves**	Zn	**Zinc (Zn):** Leaf becomes narrow and small, Lamina become chloritic, and veins remain green. Subsequently, dead spots develop all over the leaf including veins, tips and margins.
4. **Terminal bud**	Ca, B	**Calcium (Ca):** Bud leaf become chlorotic while the base remain green. Death of terminal buds in extreme cases; stem structure is weak; premature shedding of fruit and buds

	Boron (B): Yellowing or chlorosis which starts from the base to tip. The tip become elongated into whip like structure and become brownish or blackish brown; young leaves become thick, leathery, and chlorotic. Failure to set seed; internal breakdown of fruit or vegetable; death of apical buds, giving rise to witches broom

Precautions in identifying nutrient stress symptoms

1. Many symptoms appear similar. For instance, nitrogen (N) and sulfur (S) deficiency symptoms can be very alike, depending upon placement, growth stage, and severity of deficiencies.
2. Multiple deficiencies and/or toxicities can occur at the same time. More than one deficiency or toxicity can produce symptoms, or possibly a deficiency of one nutrient can induce the excessiveness of another (i.e., excessive P causing Zn deficiency).
3. Crop species, and even some cultivars of the same species, differ in their ability to adapt to nutrient deficiencies and toxicities. For example, corn is typically more sensitive to a Zn deficiency than barley (*NM 7*).
4. Pseudo (false) deficiency symptoms (visual symptoms appearing similar to nutrient deficiency symptoms). Potential factors causing pseudo deficiency include, but are not limited to, disease, drought, excess water, genetic abnormalities, herbicide and pesticide residues, insects, and soil compaction.
5. Hidden hunger: Plants may be nutrient deficient without showing visual symptoms.
6. Field symptoms appear different than 'ideal' symptoms.

Functions, Deficiency and Toxicity symptoms of the 15 Essential nutrients

Table : Generalized symptoms of plant nutrient deficiency or excess

Plant Nutrient	Type	Visual symptoms
Macronutrients		
Nitrogen	Functions	Nitrogen is important to a plant as petrol is to a car. Nitrogen constitute about 1-4 % of dry weight. It is responsible for rapid foliage growth and green color and required the production of proteins, nucleic acids (DNA and RNA), and chlorophyll. Helps plants with rapid growth, increasing seed and fruit production and improving the quality of leaf and forage crops.
	Deficiency	Chlorosis (yellowing first on lower/older leaves) In severe cases; stunted growth reduced tillering in cereals.
	Excess	Succulent growth, leaves are dark green, thick and brittle; poor fruit set; susceptible to lodging disease and insect invasion.
Phosphorus	Functions	Phosphorus is to a plant as what wheels are to a car. Concentration of phosphorus in plants is 0.1-4%. It is necessary for seed germination, photosynthesis, protein formation, flower and fruit formation. It also improve nodulation and increases water use efficiency by promoting root growth.
	Deficiency	Leaves may develop purple colouration, new leaves become dark green. Plant remain stunted, thin and weak with poor root system.
	Excess	Excess phosphorus may cause micronutrient deficiencies especially iron or zinc, necrosis, interveinal chlorosis in young leaves.
Potassium	Functions	Potassium is to a plant what brakes is to a car. Potassium content in plants varies from 1-4% on dry weight basis. It helps in the building of protein, photosynthesis and fruit quality. It helps to adjust water balance, improves stem rigidity and cold hardiness, increases disease resistance.
	Deficiency	K deficiency does not immediately result in visible symptoms (hidden hunger); Reduced growth; shortened internodes; margins of older leaves become chlorotic and burn; necrotic (dead) spots on older leaves; irregular fruit development.

Contd....

	Excess	Excess potassium may cause deficiencies in Mg, Mn, Zn and Fe.
Calcium	Functions	Calcium is to a plant what cement to a brick wall. Calcium in plants ranges from 0.2-1.0 %. It is a structural component of cell walls, influences water movement in cells and is necessary for cell growth and division. It is involved in nitrogen metabolism, reduces plant respiration and increases fruit set.
	Deficiency	New leaves becomes white , failure of terminal bud and root tips, poor fruit development and appearance.
	Excess	Excess calcium may cause deficiency in either magnesium or potassium.
Magnesium	Functions	Magnesium is to a chlorophyll what iron is to haemoglobin in the blood. Magnesium content varies from 0.2-0.4%. It is part of the chlorophyll in all green plants and essential for photosynthesis. It is necessary for functioning of plant enzymes to produce carbohydrates, sugars and fats and increases utilization of iron and phosphorus in plants. It is used for fruit and nut formation and essential for germination of seeds.
	Deficiency	Reduction in growth; yellowish, bronze, or reddish color of older leaves, while veins remains green; leaf margins may curl downward or upward with a puckering effect.
	Excess	Plant can tolerate high concentration of magnesium, but imbalance with calcium and potassium may reduce growth.
Sulfur	Functions	Master nutrient for oil production. It is a structural component of amino acids, proteins, vitamins and enzymes and is essential to produce chlorophyll. It improves root growth and seed production. It also promotes nodule formation in legumes.
	Deficiency	Chlorosis, similar to symptoms of nitrogen deficiency but occurs on young leaves.
	Excess	Excess of sulfur may cause premature dropping of leaves.
Micronutrients		
Iron	Functions	It is necessary for many enzymatic functions and as a catalyst for the synthesis of chlorophyll. Involved in redox reactions of respiration, photosynthesis.

Contd....

	Deficiency	Interveinal chlorosis of young leaves to leading to spots of dead leaf tissue.
	Excess	Entire leaf turns purplish brown.
Manganese	Functions	It is involved in enzyme activity for photosynthesis, respiration, and nitrogen metabolism. It functions as a part of certain enzyme systems and also aids in chlorophyll synthesis. It increases the availability of Phosphorus.
	Deficiency	Interveinal yellowing or mottling of young leaves.
	Excess	Older leaves have brown spots surrounded by a chlorotic circle or zone.
Zinc	Functions	It is a component of enzymes or a functional cofactor of a large number of enzymes including auxins (plant growth hormones). It is essential in carbohydrate metabolism, starch formation, protein synthesis, chlorophyll production and internodal elongation (stem growth). It also helps in seed formation.
	Deficiency	Interveinal yellowing on young leaves; little leaf and clustering of leaves at top of fruit trees due to reduced internodal.
	Excess	Excess zinc may cause iron deficiency in some plants.
Boron	Functions	Necessary for cell wall formation elongation. Aids production of sugar and carbohydrates. Essential for seed and fruit development.
	Deficiency	Death of growing points and deformation of leaves with areas of discoloration.
	Excess	Interveinal chlorosis.
Copper	Functions	It play major role in photosynthesis, reproductive stages and also indirect role in chlorophyll production. It increases sugar content, intensifies color and improves flavor of fruits and vegetables.
	Deficiency	Leaves may show interveinal chlorosis while tips of older leaves remain green.
	Excess	Fe deficiency may occur.
Molybdenum	Functions	It is a structural component of the enzyme that reduces nitrates to ammonia in plants. Without it, the synthesis of proteins is blocked and plant growth ceases. Root nodule (nitrogen fixing) bacteria also require it.

Contd....

	Deficiency	Interveinal chlorosis in older leaves; twisted leaves (whiptail); marginal scorching and rolling or cupping of leaves; nitrogen deficiency symptoms may develop.
	Excess	Intense yellow or purple color in leaves; rarely observed.
Chlorine	Function	It is involved in osmosis (movement of water or solutes in cells), the ionic balance necessary for plants to take up mineral elements and in photosynthesis.
	Deficiency	Wilted leaves which become bronze then chlorotic then die; club roots.
Cobalt	Function	Cobalt is required for nitrogen fixation in legumes and in root nodules of non legumes. The demand for cobalt is much higher for nitrogen fixation than for ammonium nutrition.
	Deficiency	Little is known about its deficiency or toxicity symptoms.
Nickel	Function	Nickel is required for iron absorption. It is essential for seed development. Plants may fail to produce viable seeds.

Chapter 6

Fertilizers and Manures

Plant requires food or nutrients or elements for its growth and development which are absorbed through soil. Adequate supply of plant nutrients is important to ensure efficient crop production. Plant nutrients are essentially applied through manures and fertilizers. Application of manures and fertilizers to the soil is one of the important factors which help in increasing the crop yield and to maintain the soil fertility.

Manures

Manures are plant and animal wastes that are used as source of plant nutrients. They release nutrient after their decomposition. These are relatively bulky materials, such as animal or green manures, which are added mainly to improve the physical condition of the soil, to replenish and keep up its humus status, to maintain the optimum conditions for the activities of soil micro-organisms and make good a small part of the plant nutrients removed by crops or otherwise lost through leaching and soil erosion.

Fertilizers

Fertilizers are industrially manufactured chemicals containing plant nutrients or it is an artificial product containing the plant nutrients which when added to soil makes it productive and promotes plant growth. Fertilizers are inorganic/synthetic substances containing one or more plant nutrients in easily soluble and quickly available form. Because of their availability in concentrated form they have the advantage of being small for bulk storage and transport and handling are easier. In addition, crops may be supplied with exact quantity of nutrient required for their growth and development in soils of varying fertility levels. Nutrient content is higher in fertilizers than in organic manures and nutrients are released almost immediately.

Difference between Manures and Fertilizers

S.no.	Characteristics	Manures	Fertilizers
1.	Nature	Plant or animal origin chemicals.	Industrially manufactured.
2.	Origin	Organic in nature.	Inorganic in nature.
3.	Type	Natural.	Artificial.
4.	Type of material	Supply organic matter.	Supply inorganic matter.
5	Nutrient availability	Slowly available.	May or may not be readily available.
6.	Concentration of nutrients	Less concentrated.	More concentrated.
7.	Effect on soil	Improves structure, water holding capacity, permeability of soil.	Have no effects in soil.
8.	Supply of nutrients	Supply all the primary nutrients including micronutrient.	Supply specific type of nutrients one, two or three or different micro nutrients.
9.	Acidity/Alkalinity	Manures does not produce acidity or alkalinity in soil.	Fertilizers produces acidity (Ammonium sulfate, urea etc.) and alkanility (Sodium nitrate in soil).

Classification of Manures

A. Bulky Organic Manures

They contain low amount of plant nutrients and are applied in large quantities.

1. Farm yard manure (F.Y.M.)
   a. Cattle manure
   b. Poultry manure
   c. Sheep and goat manure
2. Compost
   a. Farm compost
   b. Town compost
   c. Vermicompost
   d. AZO compost
   e. Super compost

B. Green Manures

a. Leguminous plant
b. Non-leguminous plant

C. Concentrated Manure

They contain higher percentage of major plant nutrients. *e.g.* various oil cakes and waste products of animal origin like dried blood, bone meal, fish manure, etc.

1. **Oil cakes:** Richest source of plant nutrients of all organic manures.
   a. Edible oil cakes: (Used for cattle feeding).
      Mustard cake, Groundnut cake, Sesame cake, linseed cake etc.
   b. Non edible oil cakes (Used as manures).
      Castor cake, mahua cake, neem cake, sunflower cake, karanja cake.
2. **Waste Products from slaughter house**
   a. Blood meal.
   b. Bone meal
3. **Fish products:** fish meal etc.

D. Guano: Material obtained from the excreta and dead bodies of sea bird.

E. Biofertilizers

a. Rhizobium
b. Azotobacter
c. Azospirillum
d. Blue green algae (BGA)
e. Azolla
f. Mycorrhizae

Effect of bulky organic manures on soils is three fold

- Have direct effect on plant growth as they contain both major and micronutrients.
- Increase organic matter and humus content, improve the physical properties of soils and water holding capacity of soil.
- Provide food for soil microorganisms. This increases activity of microbes which in turn help convert unavailable plant nutrients into available forms.

1. Farm-Yard Manure: It is the most commonly used organic manure in India. It is also called barn manure, stable manure, dung and cattle manure. It is generally applied at the rate of 10-20 t/ha in the field about 15-20 days before sowing. About 30% N, 60-70 % P_2O_5 and 70 % K_2O are available to the first crop.

a. **Cattle manure:** Cattle manure contains 0.5-1.0 % N, 0.2 % P_2O_5 and 0.5 % K_2O. It consists of a mixture of cattle dung, the bedding, used in the stable and of any ruminants of straw and plant stalks fed to cattle. Farm-yard manure refers to the decomposed mixture of dung and urine and farm animals along with the litter (bedding materials) and left over material from roughages and fodder fed to cattle. Farm-yard manure consists of two original components – the solids or dung and liquid or urine. Urine of all animals contains more percentage of nitrogen and potash, compared to the dung portion.

Trench method of preparing FYM: This method has been recommended by C.N. Acharya to avoid nutrient loss. In this method, the waste material of cattle shed consisting of dung and urine soaked in the refuse of the shed is collected daily and placed in trenches about 6-7.5 metre long, 1.5-2 metre broad and 1 metre deep. Each trench is filled to a height of 45-60 cm in a dome-shaped and is plastered over with cowdung with slurry. Thus, the material is allowed to get decomposed and it is ready in about 3-4 months.

b. **Sheep and goat manure:** It contains about 1.5% N, 0.6 to 1% P_2O_5 and 2.0% K_2O. It is also valuable organic manure. Sweeping of sheep and goat sheds are placed in pits for decomposition and it is applied later to the field. It is estimated that by stocking 1000 sheep for a night will add about 2 tons of droppings.

c. **Poultry manure:** The excreta of birds contains both solid and liquid parts and hence there is no urine loss. The decomposition rate of the poultry manure is very quick. It contains about 3% N, 2.6% P_2O_5 and 1.0% K_2O.

2. Compost: Compost is well rotten organic manure prepared by decomposition of organic matter while composting is a biological process in which microorganisms (aerobic or anaerobic) decompose the organic matter, thereby lowering down the C:N ratio.

Method: A trench of 5 meter long, 1.6 metre wide and 1 metre deep is made and waste material is spread in the trench in 30 cm layer. Sprinkling of slurry of cowdung and water is done, then comes the layer with urine, earth and cowdung slurry. This process is repeated till the height reaches to 50cm. The top is covered by thin plaster of earth and dung and it is left undisturbed for 3-4 months. Then, it is ready to apply in fields.

Different methods of composting

a. NADEP method: It was started by farmer (Narayana Dev Rao Panthary Pande) of Pusad village in Maharashtra for composting when availability of cow dung is low.

b. Indore method

c. Super digested compost

d. Banglore process / Country method

AZO Compost: Compost prepared from Nitrogen fixing bacteria is called AZO compost. It is the cheapest source of N among all organic manures. It contains 1.5% N, 0.9% P_2O_5 and 0.8% K_2O.

There are different types of composts

a. **Farm Compost:** It contain 0.5% N, 0.15% P_2O_5 and 0.5% K_2O made from paddy straw, sugarcane fresh weeds and other waste.

b. **Town Compost:** It contain 1.4% N, 1% P_2O_5 and 1.4% K_2O and made from town refuses like night soil, dustbin refuse and street sweepings.

c. **Vermicomposting:** Vermicomposting refers to the controlled degradation, or composting, of organic wastes, primarily by earthworm consumption. Vermiculture means scientific method of breeding and raising earthworms in controlled condition. Vermi technology is the combination of vermiculture and vermicomposting. Generally there are 3200 species of earthworms occurring in nature. Out of these *Eisenia fetida* and *Eudrelis eugina* are used for preparing vermicompost.

Advantages

- Improve soil structure and enrich the nutrient status of soil.
- Earthworms churns the soil and make it porous.
- Improve the water infilteration rate.
- Neutralize soil pH.
- Vermicompost is ready in 2-2.5 months. Vermicompost is black, light weights and has no bad smell.

d. *Super Compost:* When super phosphate is used during composting.

B. Green Manure

It is the practice for enriching the soil by ploughing or turning into the soil undecomposed green plant tissues for the purpose of improving structure and fertility of the soil. On an average, it can fix 60-80 kg/ha of N. There are two methods of green manuring

1. Green manuring *in-situ*

2. Green leaf manuring

1. GM *In-situ*: Growing of crop and incorporation in the same field with the soil is called *in-situ* method.

Leguminous green manure crops: Dhaincha (*Sesbania acueleata*), Sunhemp (*Crotalaria juncea*), Dhaincha (*Sesbania rostrata*), Indigo (*Indigofera tinctoria*) and Khesari (*Lathyrus sativus*).

Non leguminous green manure crops: Bhang (*Cannabis sativa)*, Kodogira (*Vernonia cenerea*).

2. Green leaf manuring: Green leaves are incorporated. eg. Glircidia (*Glyricidia maculate*), Karanj (*Pongamia pinnata*), Ipomoea (*Ipomoea carnea*).

Green manuring crop should be buried in the soil when it reaches the flowering stage. Sunhemp is ready for turning after 7-8 weeks while dhaincha at 4-8 weeks. Legume crops, leaves of bushes and trees is buried in the soil and allowed to decompose in the field.

Methods

Plants are cut close to the ground and green material is put in the furrows opened by mould board plough and later buried. Material is planked down with a heavy plank or log, and then field is ploughed. Mix the uprooted plants by disc harrow.

Advantages

1. Increase organic matter and nutrient availability.
2. Checks weed growth.
3. Reclamation of saline and alkali soils.
4. Improve water holding capacity and soil health.

Table: Amount of N content of green manure crops

Crop	Botanical name season	Growing green	Average yield of matter (q/ha)	Nitrogen added in the soil (kg/ha)
Dhaincha	*Sesbania aculeate*	Kharif	144	77.1
Sannhemp	*Crotalaria juncea*	Kharif	152	84.0
Moong	*Phaseolus aureus*	Kharif	57	38.6
Cowpea	*Vigna sinensis*	Kharif	108	56.3
Guar	*Cymopsis tetregonoloba*	Kharif	144	62.3
Pea	*Pisum sativum*	Rabi	200	80
Berseem	*Trifolium alexandrium*	Rabi	111	60.7

C. Concentrated Manure

They contain higher percentage of major plant nutrients. e.g. various oil cakes and waste products of animal origin like dried blood, bone meal, fish manure, etc.These are also known as organic nitrogen fertilizer.

1. Oil cakes: Richest source of plant nutrients of all organic manures. After oil is extracted from oil seeds, the remaining solid portion is dried as cake which can be used as manure. There are many varieties of oil cakes which contains not only nitrogen but also some P and K along with large percentage of organic matter. These oil cakes are of two types.

a. Edible oil cakes: (Used for cattle feeding)
   Mustard cake, Groundnut cake, Sesame cake, Linseed cake etc.

b. Non edible oil cakes (Used as manures)
   Castor cake, mahua cake, neem cake, sunflower cake, karanja cake.

Oil cakes are quick acting organic manure. Though they are insoluble in water, their nitrogen become quickly available to plants in about a week or in 10 days after application. Oil cakes should be well powdered before application, so that they can be spread evenly and are easily decomposed by microorganisms. Depending on crops, oil cakes are applied as broadcast, drilled or placed near root zone while earthing up. The average nutrient composition of oil cakes are as under:

Table: Average percentage composition in different oil cakes

Sl.No.	Organic manure	N	P_2O_5	K_2O
1	Groundnut cake	7.29	1.53	1.33
2	Linseed cake	5.56	1.44	1.28
3	Castor cake	4.37	1.85	1.39
4	Neem cake	5.22	1.08	1.48

2. Waste Products from slaughter house

a. **Blood meal:** Dried blood or blood meal is bi-product of slaughter house. It contains 10-12% nitrogen, 1-1.5% phosphorus and 1.0% potassium. It is very quick acting manure and is effective on all crops and on all types of soils.

b. **Bone meal:** Bones from slaughter houses, carcasses of all animals and from meal industry constitute bone meal, which is the oldest phosphatic fertilizer used. It also contains some N. Various types of bone meal are:

   i. Raw or untreated bone meal: Bones collected from city slaughter houses and from countryside are dried and powdered without any treatment. It contains 2.4% N and 20.0 to 25.0% P_2O_5.

   ii. Steamed bone meal : Bones are soaked in water and caustic soda. They are sterilized with steam and then they are dried in warm rotating oven and powdered. It contains about 27.0% P_2O_5 and 1.0% N. It is well suited fertilizer for acidic soil and also for long duration crops.

Fish products: fish meal etc.

Fish meal: Fish manure or meal is processed by drying non-edible fish. Carcasses of fish and wastes from fish industry. It contains 4.0-10.0% nitrogen, 3.0-9.0% P_2O_5 and 0.3 to 1.5% K_2O. It has an offensive smell. It is available either as a dried fish or as fish meal or powder. The fish is dried, crushed or powdered and filled in bags. Fishmeal is quick acting organic manure and is suitable for application to all crops on all soils.

D. Guano

Material obtained from the excreta and dead bodies of sea bird. It contains 7-8 % Nitrogen, 11-14% phosphorus and 2.3-3.0% potassium. It is very quick acting manure and is effective on all crops and on all types of soils. It is applied few days prior to sowing or at sowing time or as a top dressing .

E. Biofertilizers

The term "biofertilizers" include selective microorganisms like bacteria, fungi and algae which are capable of fixing atmospheric nitrogen or convert insoluble phosphates in the soil into forms available to the plants.

Saprophytes

Microorganisms that are capable of decomposing organic matter at a faster rate can be used as a fertilizer for quick release of nutrient. Aspergillus,

Penicillium, Trichoderma are celluloytic fungi which bread down cellulose of plant material. The natural process of decomposition is accelerated and composting time is reduced by 4 to 6 weeks by the use of inoculants of these organisms.

Symbiotic Bacteria

Bacteria belonging to the genus *Rhizobium* are capable of fixing atmospheric nitrogen in association with leguminous crops. Different species of *Rhizobium* are used for treating the leguminous crops (Table).

Rhizobium species enter the roots of host plants and form nodules on the root surface. Nitrogen fixed by the Rhizobium is translocated through xylem vessels of the host plant mainly in the form of aspergine and to some extent as gluatamine. This inoculum can be applied in three ways Ist Seed Treatment, IInd Soil Treatment, IIIrd Soil Application.

Among them, seed treatment is the best.

Table : Rhizobium species suitable for different crops.

R. leguminosarum	Peas (*Pisum*), lathyrus, vicia, lentil (*Lens*)
R. trifolii	Berseem (*trifolium*)
R. phaseoli	Kidney bean (*phaseolus*)
R. lupine	Lupinus, Ornithopus
R. japonicum	Soybean (Glycine)
R. melilotii	*Melilotus*, Lucerne (*Medicago*) Fenugreek (*Trigonella*)
R. cowpea miscellany	Cowpea, clusterbean, greengram, blackgram, redgram, groundnut, moth bean, dhaincha, sunhemp, Glyricidia, Acacia, Prosopis, Dalbergia, Albizzia, *Indigofera, Tephrosia, Atylosia,* stylo
Separate group	Bengalgram (gram)

Free Living organism

The important free living organisms that can fix atmospheric nitrogen are blue green algae (BGA), *Azolla, Azotobacter* and *Azospirillum*. Among them, BGA and *Azolla* can survive only in lowland conditions.

Blue-green-Algae (BGA): Several species of blue–green-algae can fix atmospheric nitrogen. The most important species are *Anabaena* and *Nostoc*. The amount of nitrogen fixed by blue-green-algae ranges from 15 to 45 kg N/ha.

Azolla: Azolla is a free floating fresh water fern. *Azolla pinnata* is the most common species occurring in India. It fixes nitrogen due to Anabaena species of blue-green algae present in the lobes of Azolla leaves. A thick mat of Azolla supplies 30 to 40 kg N/ha.

Azotobacter and Azospirillum

Azotobacter chroccum is capable of fixing 20 to 30 kg N/ha. It can be applied by seed inoculation, seedling dip or by soil application.

Mycorrhiza and Phospho – Microorganisms: Phosphorus availability and fertilizer phosphorus use efficiency can be increased with mycorrhiza, phosphate solubilizing bacteria and fungi. Mycorrhiza inhabits roots of several crops and solubilises soil phosphates. Inoculation of mycorrhiza increases the pod yield of groundnut. Some microorganisms like *Psuedomonas striate, Aspergillus awaneorlii and Bacillus polymyxa* are capable of solubilising phosphates. The inoculum of these microorganisms is applied to increase the availability of phosphorus.

Other types

1. **Short and long manure:** Short manure is the decomposed manure which has lost its original structure. Long manure is the fresh manure which have pieces of straw and other materials.
2. **Hot and cold manure:** Hot manure is the manure obtained from the excreta of horses and sheep while pig and cattle manure is called cold manure.
3. **Fanging:** It is profuse fungal growth on the surface of moist manure, giving it on ashy appearance.
4. **Night soil poudrettes:** Night soil is human excreta, both solid and liquid. Night soil contains appreciable amount of nitrogen. It can be profitably coverted into useful manure known as 'POUDRETTE' Dehydration of night soil can be done as such, or admixture of absorbing materials like soil, ash, charcoal, and saw dust. Among the materials saw dust is the most effective one which can be very well used. Cow dust has high dehydrating capacity as well as the property of absorbing foul smell. About 40-45% of saw dust could be added with the night soil. By this way, it is possible to get a dry, acidic poudrette which may contain 2-3% N. Odourless manure can also be obtained by admixing night soil with equal volume of ash and 10% of powdered charcoal. The poudrette produced by this way contains 1.3% N, 2.8% P_2O_5, 4.1% K_2O and 24.2% lime.

Fertilizers

Fertilizer is any organic or inorganic material of natural or synthetic origin (other than liming materials) that is added to a soil to supply one or more plant nutrients essential to the growth of plants OR Any commercial chemical which is added in the soil to boost up the yield of the crop.

Classification of Fertilizers

A.Straight Fertilizers

a. Nitrogenous fertilizers

i. Ammonical fertilizers (NH_4 fertilizers)

ii. Nitrate fertilizers (NO_3 fertilizers)

iii. Nitrate and ammonium fertilizers

iv. Amide fertilizers

b. Phosphatic fertilizers

i. Water soluble phosphatic fertilizers

ii. Citrate soluble (water insoluble) phosphatic fertilizers

iii. Water and citrate insoluble phosphatic fertilizers

c. Potassic fertilizers

B. Complex/compound Fertilizer

a. Complete fertilizers

b. Incomplete fertilizers

C. Mixed fertilizers/fertilizers mixture

D. Micronutrient fertilizers

These are those fertilizers having micronutrients *e.g.* zinc sulfate, magnesium sulfate etc.

I. Based on nutrients constitution, fertilizers are grouped as

1. Nitrogenous fertilizers
2. Phosphatic fertilizers
3. Potassic fertilizers

II. Based on physical form, fertilisers can be grouped as

1. Solid fertilizers
2. Liquid fertilizers

Solid fertilizers are in several forms *viz.*

1. Powder (single superphosphate)
2. Crystals (ammonium sulphate)
3. Prills (urea, diammonium phosphate, superphosphate)
4. Granules (Holland granules)
5. Supergranules (urea supergranules)
6. Briquettes (urea briquettes)

Liquid Fertilizers

1. Liquid form fertilizers are applied with irrigation water or for direct application.
2. Ease of handling, less labour requirement and possibility of mixing with herbicides has made the liquid fertilisers more acceptable to farmers.

Types of fertilizers

a. **Straight fertilizers:** It contains only one primary nutrient element e.g. urea, ammonium sulfate.It is further divided into nitrogenous, phosphatic and potassic fertilizers.

b. **Complex/compound fertilizer:** The fertilizers having two or more primary essential materials (N,P,K) are called complex fertilizer. E.g. Diammonium Phosphate. NPK complex fertilizers. These are of two types:

   i. Complete fertilizer: It contain three primary major nutrients N,P.& K.

   ii. Incomplete fertilizers: It contain any two primary nutrients.

c. **Mixed fertilizers/fertilizers mixture:** A mixture of two or more straight fertilizer material is known as mixed fertilizer.

d. **Micronutrient fertilizers:** These are those fertilizers having micronutrients *eg.* zinc sulfate, magnesium sulfate etc.

Other Types

1. **Binary fertilizers:** It contains two major nutrients e.g. potassium nitrate.
2. **Ternary fertilizers:** It contains three major nutrients e.g. Ammonium Potassium Phosphate.

3. **Low analysis fertilizers:** They have less than 25% of primary nutrients *eg.* SSP, Sodium nitrate.

4. **High analysis fertilizer:** Having more than 25% of primary nutrients content *eg.* urea, Anhydrous ammonia, DAP.

Straight Fertilizers

Nitrogeneous fertilizers: According to the manner in which their nitrogen is combined with other elements, the nitrogenous fertilizers are divided into four groups; nitrate, ammonia and ammonium salts, chemical compounds containing nitrogen in the amide form, and plant and animal by-products.

Type of fertilizer	Form or nitrogen	Nature of Fertilizer	Examples
Nitrogenous Fertilizer			
1. Nitrate fertilizer	NO_3^-	1. Highly reactive and mobile. 2. Are basic in their residual effect in soils 3. Susceptible to losses under water logged condition. 4. Effective in most of the crops except paddy.	a. Sodium Nitrate (16% N) b. Calcium Nitrate (15.5% N) c. Potassium nitrate (13 %N)
2. Ammonium fertilizer	NH_4^+	1. Soluble is water 2. Are acidic in their residual effect in soils 3. Adsorbed on clay, therefore less susceptible to loses 4. Effective in paddy	a. Ammonium sulphate – 20.6%N b. Ammonium chloride – 25% N c. Ammonium phosphate-20% N d. Anhydrous ammonia – 82%N e. Ammonia solution-20-25% N
3. Nitrate and Ammonium fertilizer	NO_3^-, NH_4^+	1. Soluble in water 2. Acidic in nature	a. Ammonium nitrate – 33-34%N b. Calcium Ammonium nitrate– 25% N c. Ammonium sulfate nitrogen - 26% N
Organic Nitrogenous Fertilizer			
4. Amide fertilizer	(NH_2)	Not directly available to plants, quickly converted by soil microbes to ammonia and nitrate form	a. Urea 46% N b. Calcium Cyanamide 21% N
5. Slow release N fertilizer		Release nitrogen very slowly	a. Urea formaldehyde 38% N

Contd......

b. Oxamide– 31.8% N
c. Isobutylene diurea (IBDU) – 32.2%
d. Crotonilidine diurea (CDU) – 32% N
e. Guanyl urea (GU) – 37% N
f. N-lignin – 18% N
g. Sulfur coated urea 36-40%N
h. Metal Ammonium phosphate
If metal is Mg–8.3%N
If metal is Fe–7.5% N
If metal is Cu–7.2%N
If metalis Zn–7.8% N
If metal is Mn–7.5%N
If metal is Co– 6.1 %N

Phosphatic fertilizers: According to the solubilities, the phosphatic fertilizers are divided in following groups.

Water soluble: These are used for quick start and for short duration crops like wheat, for neutral and alkaline soils e.g.

Single superphosphate (SSP): 16-20 % P_2O_5
Double superphosphate (DSP): 32 % P_2O_5
Triple superphosphate (TSP): 46-48 % P_2O_5
Monoammonium phosphate (MAP): 12 % N and 48 % P_2O_5
Diammonium phosphate (DAP): 16-48-0; 18-46-0 (N-P_2O_5 - K_2O)

Citrate soluble (water insoluble): These are used in acid soils and for long duration crops like sugarcane, lowland rice, tapioca and tea.

Dicalcium phosphate: 14 % P_2O_5
Basic slag: 20 % P_2O_5
Calcium metaphosphate: 60-64 % P_2O_5

Citrate and water insoluble: These are used in strongly acidic soils and for plantation crops.

Rock phosphate: 20-30 % P_2O_5
Bonemeal steamed: 1-2 % N and 20-30 % P_2O_5
Raw bonemeal: 3-4 % N and 20-25 % P_2O_5

Other Phosphatic fertilizers

Kotka phosphate: 26% P_2O_5
Rhenmia phosphate : 26-28% P_2O_5

Thermophos: 18.3 % P_2O_5
Pelophos: 18 % P_2O_5
Coronat phosphate: 21% P_2O_5

Potassic fertilizers

Potassium chloride / muriate of potash (MOP): 60% K_2O
Potassium sulphate/ sulphate of potash: 48% K_2O
Potassium nitrate/ nitrate of potash: 13% N and 44% K_2O

Complex multiple nutrient fertilizers: The fertilizers which contain two or more primary essential nutrients (N,P,K) are called complex fertilizers. The complex fertilizers containing two primary nutrients are called incomplete while those containing three nutrients called complete complex fertilizers. They are much more desirable for balanced treatment of soil than the straight fertilizers. These fertilizers are generally granular and free flowing. Therefore ensure uniform distribution in the field while application. Three types of complex fertilizers are produced in India.

a. **Ammonium phosphates:** Produced by neutralizing ammonia with phosphoric acid enriched by nitrogenous fertilizers. The following fertilizers are produced.

Monoammonium phosphate	1-48-0
Ammonium phosphate sulphate	16-20-0, 20-20-0
Urea ammonium phosphate	24-24-0, 28-28-0
Diammonium phosphate	18-46-0

b. **Nitrophosphates:** These fertilizers are manufactured by treating rock phosphate with nitric acid or mixture of acids. Following commonly used nitrophosphates are produced

Nitro carbonic process (16-14-0), PEC nitrophosphate
Seperation process ODDA nitrophosphate
Sulphonitric nitrophosphate 12.9-12.9-0

c. **N.P.K. complex fertilizers** – Produced by mixing ammonium phosphate or nitrophosphate and potasic fertilizers. Examples are NPK 18-18-9, 15-15-15-19-19-19-, 11-22-22, 10-26-26, 12-32-16.

Advantages of complex fertilizers

1. The possibility of adulteration is generally less
2. Each granule is homogenous in nutrient content
3. Being granular, the drilling of fertilizers is easy

4. They are cheaper than straight or mixed fertilizers
5. Phosphorus availability is not affected (fixation is less) as the phosphorus in the granules will have less contact with soil particles.

Disadvantages of complex fertilizers

The disadvantage with Complex fertilzer is that the ratios of the nutrients are fixed and the farmers may have to supplement with straight fertilizers to meet the crop requirements.

Mixed fertilizers: A mixture of two or more straight fertilizer material is referred to as mixed fertilizer or fertilizer mixture. Mixed fertilizers are physical mixtures of fertilizer materials containing two or three major plant nutrients. Mixed fertilizers are made by thoroughly mixing the ingredients either mechanically or manually.

Advantages

1. Less labour is required to apply a mixture than to apply various components separately. This is an important factor in areas where farm labour is scare and expensive.
2. If a proper mixture is used to suit a particular soil type and crop, use of a fertilizer mixture leads to balanced manuring. This gives higher yield and more profit to the cultivators.
3. The residual acidity of fertilizers can be effectively controlled by the use of a proper quantity of liming materials in the mixtures.
4. Micronutrients which are applied in small amounts to soil can be incorporated in fertilizer mixtures. This facilitates uniform soil application of plant nutrients required in small quantities.
5. Mixtures have a better physical condition (granulated) and are more easily applied than many straight fertilizers.

Micronutrient Fertilizers

Sixteen elements are essential for plant growth out of these Fe, Zn, Mn, Cu, B, Mo, Cl are required in small quantities. They are called tertiary or micronutrient. The average concentration of these nutrient in soil are Mn 1000 ppm, Cl 480 ppm, Zn 80 ppm, Cu 70 ppm, B 10 ppm, Mo 2 to 4 ppm, iron 140 ppm.

The deficiencies may be overcome by addition of more than one micronutrients. If more than one micronutrients are applied it is called multinutrients. The information of sources of micronutrients. Their formula and nutrient content is given in following table.

Nutrient contents of micronutrient fertilizer

S.No.	Name	Formula	Element/forms	Contents (%)
1.	Zinc Sulphate*	$ZnSO_4$, $7H_2O$	Zn	21.0
2.	Mangnese Sulphate*	$MnSo_4$	Mn	30.5
3.	Ammonium Molybdate	$(NH_4)_6$ $Mo_7O_{24}4H_2O$	Mo	54.0
4.	Borax (for soil application)	$Na_{-2}B_4O_7,5H_2O$	B	10.5
5.	Solubor (Foilar spray)	$Na_2B_4O_7,5H_2O+$ $Na_2B_{10}O_{16},10H_2O$	B	19.0
6.	Copper Sulphate*	$CuSO_4,5H_2O$	Cu	24.0
7.	Ferrous Sulphate*	$FeSO_4,7H_2O$	Fe	19.5
8.	Chelated Zn	As Zn=EDTA	Zn	12.0
9.	Chelated Fe	As Fe-EDTA	Fe	12.0
10.	Zinc Sulphate monohydrate	$ZnSO_4,H_2O$	Zn	33.0

*Average Sulphate Content

i) $ZnSO_4$ = 15%
ii) $MnSO_4$ = 17%
iii) $CuSO_4$ = 13%
iv) $FeSO_4$ = 19%

Secondary fertilizers

S.No.	Fertilizer	Ca%	Mg%	S%	Others%
1.	Gypsum ($CaSO_4.2H_2O$)	29.2	—	18.6	-
2.	Rock phosphate	33.1	—	—	25.2%P_2O_5
3.	SSP	19.5	—	12.5	16%P_2O_5
4.	$MgSO_4.7H_2O$ (Epsom salt)	—	9.6	13.0	—
5.	Potassium sulphate	—	—	17.5	48 % K_2O
6.	Ammonium sulphate	—	—	24.2	21% N
7.	Basic slag	—	—	3.0	15. 6%P_2O_5
8.	$CuSO_4$	—	—	11.4	21 %Cu
9.	Fe SO_4	—	—	18.8	32.8 %Fe
10.	$ZnSO_4$	—	—	17.8	36.4 %Zn
11.	FeS_2 (Iron pyrite)	—	—	22-24	—
12.	Borax ($Na_2B_4O_7.10H_2O$)	—	—	—	10.6 % B

Nitrification Inhibitors and slow release fertilizers

These are used to reduce the leaching, volatilization and denitrification losses, so as to increase the nutrient availability. They release the nitrogen slowly and uniformly.

Nitrification Inhibitors

For upland: AM (2-amino-4 chloro-6-methyl pyrimidine); N-serve(2-chloro-6-trichloromethyl pyridine)

For lowlands: Oxamide, Dicyandiamide, Thiourea, Urea pyrolyzate

Others: Neemcake, Nitrapyrin, Guanyl thiourea, sulphathiazole.

Slow release fertilizers

For reducing the leaching losses, the solubility of nitrogen fertilizers are reduced by:

- *Coating Barriers:* Neem coated urea, Sulphur coated urea, Lac coated or shellac coated urea.
- *Use of Super granules:* Gromor (Urea ammonium phosphate), Ammophos-B (Ammonium phosphate sulphate).
- *Synthesizing compounds which are less soluble:* IBDU (Isobutylidene diurea), CDU (Crotonylidene diurea), GUS (Guanyl diurea), Urea formaldehyde, Oxamide.

Methods of Fertilizer Application

A. Application in solid form

B. Application in liquid form

Application in solid form

1. Broadcasting
2. Placement
3. Localized placement

Broadcasting: Application of fertilizer uniformly on the soil surface is known as broadcasting. This is done either before sowing of the crop or in the standing crop. Broadcasting is the most widely practical method in India due to ease in application. It is advantageous with solid and soluble fertilizers.

1. *Basal application:* Evently spreading of solid fertilizers over the entire field before or at the sowing or planting. Generally the entire dose of phosphatic and potassium fertilizers are applied by broadcasting before sowing. Because of their low mobility in soil these fertilizers are incorporated in the root zone.
2. *Top dressing:* The broadcasting of the fertilizer on closed sown standing crops. Generally nitrogen fertilizers are used for top dressing.

Placement

1. *Plough sole placement:* The fertilizer is placed in a continuous band on the bottom of furrow during the process of ploughing.
2. *Deep placement :* Usually nitrogenous fertilizer (urea) is placed in reduction zone (deep in soil), where it remains in ammonial form and is available to the crop slowly.
3. *Sub soil placement:* It refers to the placement of fertilizers in the sub soil with the help of high power machinery. This method is recommended in humid and subhumid regions where sub soils are acidic. Fertilizers like P and K are placed in subsoil.

Localized placement: Fertilizer are applied close to the seed or plant in such a way that it has least contact with the soil and is available to plants.

1. *Band Placement:* Application of fertilizers in narrow bands beneath and by the side of the crop rows. It is done under the following situations:
    - When small quantities of fertilizers are to be distributed.
    - When phosphatic fertilizers are applied in acidic soils where fixation of phosphorus is a problem.
    - In the case of crops sown in wide rows.
    - In the case of shallow rooted plants.
    - On soils with low fertility.

Depending upon the root system fertilizer band can be 5cm below the seed. In cereals and millets, which produce fibrous roots, it is advantageous to place fertilizers 5cm away from the seed row and 5cm deeper than the seed placement.

2. *Point placement:* Placement of the fertilizer near the plant either in a hole or in dispersion followed by closing or covering with the soil is known as point placement of fertilizers. It is adopted for top dressing of nitrogenous fertilizers in widely spaced crops.
3. *Ball placement:* In case of maize or vegetables crops when plant spacing is more small balls of fertilizers are made and placed near the plants in the soil.
4. *Hill placement:* If the spacing is more the fertilizer is applied near the roots of the plants i.e. maize, cotton etc.

5. *Ring placement:* This is followed in orchards. Fertilizers are applied around the plants in a ring.
6. *Drill placement:* Fertilizers are drilled along with the seed using suitable equipment, *eg.* seed cum fertilizer drill. Seed cum fertilizer drill is an implement useful to sow the seed in rows as well as to apply the fertilizers at the same time. The implement is designed to place the fertilizer 5cm (2") below the seed and 5cm (2") away from the seed avoiding fertilizer injury to the seed. Separate hoppers are provided for the seed and fertilizer. There is also arrangement to regulate the seed and fertilizer rates as per the crop requirement. This is efficient way of applying fertilizer and sowing the seed. Additionally, the drill ensures uniform stand and growth of plants. Using seed cum fertilizer drill all most all crops can be sown.

Liquid form

Fertigation: Application of fertilizers with irrigation water is known as fertigation. It is generally followed with drip irrigation.

Foliar spray: This method is important in micronutrients. Foliar fertilizers are dilute solutions applied directly to the leaves. Fertilizers are dissolved in water and such diluted solutions are sprayed directly on the plants foliage. Hand operated sprayers are used for smallholdings. Tractor drawn low volume sprayers can be used while on large scale aircrafts are used for foliar spraying.

Optimum concentration for spraying micronutrients

Zinc sulfate	0.5%
Borax	0.2%
Ferrous sulfate	0.5%
Molybdic acid	0.05%
Manganese sulfate	0.2%
Copper sulfate	0.2%

Fertilizer grade : Refers to the guaranteed content (analysis) of plant nutrients in a fertilizer. For example 5-5-10 fertilizer grade means that 100 kg of mixed fertilizer will contain a minimum of 5% nitrogen (N), 5% of phosphorus (P_2O_5) and 10% of potassium (K_2O).

Balanced fertilizer application

Balanced fertilizer application refers to the practice of applying the required plant nutrients after taking into account available nutrients in the soil.

Time of application

1. Nitrogen is required throughout the crop growth. It is taken up by the plant slowly in beginning , rapidly during the grand growth and again slowly in the maturity.
2. N is lost easily by leaching. Therefore it is best to use it in split doses.
3. P is required during the early root development and early plant growth. As such crop utilize 2/3 of total requirement of P when plant accumulates 1/3 weight.
4. All phosphatic fertilizers release P for plant growth slowly. As such it is always recommended that the entire quantity of P fertilizers should be applied before sowing or planting.
5. Rate of absorption of potash is like N i.e. it is absorbed upto harvesting stage. But potash is available slowly. Therefore entire dose of K should be applied at sowing time.
6. Leaching of K is more in sandy soils and therefore should be applied in split doses in sandy soils.

Efficient use of fertilizers

1. Select the most fertilizer responsive and best suited crops and their varieties for the locality.
2. High yielding varieties (HYV) of crops give higher yield than local varieties without fertilizer application as well.
3. Balanced fertilization should be practiced based on soil test. Fertilizer recommendations should be based on the crop sequence and not on individual crop basis.
4. While all of phosphate and potash are applied as basal dressing and nitrogen should be applied in split doses.
5. Phosphate should be placed 4 to 6 cm below and 4 to 6 cm away from the seeds to ensure maximum availability.
6. To the extent possible irrigate the field with just enough water (In dry soil, fertilizers should be placed only in the moist zone.
7. Sowing of crop should be done timely depending on the locality to get the benefit of maximum efficiency of applied fertilizers.

8. Optimum plant population of the crop needs to be maintained by adopting proper plant spacing.
9. Timely control of pests and diseases will help in realizing maximum effectiveness from fertilizers.

Integrated Nutrient Management

Integrated nutrient management (INM) refers to the practice of integrated use of all natural and man-made resources of plant nutrients so that the crop productivity increases efficiently and environmentally friendly manner without sacrificing the soil productivity of the future generations.

Sufficient and balanced application of organic manures and fertilizers is the key component of INM.

How to economize fertilizer use?

- The fertilizer recommendations should be based on soil test values.
- Balanced use of fertilizer should be advocated on better economic returns.
- Use of nitrogenous fertilizer in split doses economizes fertilizer use.
- Micronutrient deficiencies should be corrected as and when needed.
- Fertilizer schedule should be adopted for the whole crop sequence instead of a single crop.
- To get the maximum benefit from the applied fertilizers crops should be irrigated at the critical growth stages.

Conversion factors determining quantities of fertilizers

Qunatity	Multiplied by	Gives corresponding quantity
Nitrogen	4.854	Ammonium sulphate
Nitrogen	2.222	Urea
Nitrogen	3.846	Ammonium sulphate nitrate
Nitrogen	4.000	Ammonium chloride
Nitrogen	3.030	Ammonium nitrate
Phosphoric acid (P_2O_5)	6.250	Superphosphate, single
Phosphoric acid (P_2O_5)	12.222	Superphosphate, double
Phosphoric acid (P_2O_5)	2.857	Dicalcium phosphate
Phosphoric acid (P_2O_5)	5.000	Bone meal, raw

Contd....

Potash (K_2O)	1.666	Muriate of potash
Potash (K_2O)	2.000	Sulphate of potash
Ammonium sulphate	.206	Nitrogen
Sodium nitrate	0.155	Nitrogen
Urea	0.450	Nitrogen
Ammonium sulphate nitrate	0.260	Nitrogen
Ammonium chloride	0.250	Nitrogen
Ammonium nitrate	0.330	Nitrogen
Superphosphate, double	0.450	Phosphoric acid (P_2O_5)
Dicalcium phosphate	0.350	Phosphoric acid (P_2O_5)
Bone meal, raw	0.200	Phosphoric acid (P_2O_5)
Uriate of potash	0.600	Potash (K_2O)
Sulphate of potash	0.500	Potash (K_2O)

Chapter 7

Irrigation and Drainage

Water is indispensable for human, animal and plant life. Water is essential part of protoplasm. It is essential for photosynthesis. About 400 to 500 litres of water is necessary for the production of a kilo of plant dry matter.

Irrigation

It is the human manipulation of hydrological cycle to improve crop production and quality and to decrease the economic effects of drought. Israelsen and Hansen (1962) defined irrigation as the artificial application of water for the purpose of supplying moisture to plant growth.

Importance of water in plants

- Water act as reagent in photosynthesis.
- Water act as a solvent.
- Water maintains turgidity.
- Water is the constituent of cell protoplasm: plant usually contains 85-90% water.

Purpose of irrigation

- To soften the soil for tillage.
- To supply moisture essential for plant growth.
- To cool the soil and atmosphere, thereby making more favourable micro-environment for plant growth.
- To provide crop insurance against short duration drought.
- To leach or dilute excessive salts in the crop root zone, thereby providing

a favourable environment in the soil profile for absorption of water and nutrients.

Soil Moisture Constants

The status of soil mass or changes occurring in the soil mass after the irrigation or rainfall is called the soil moisture constant. Different moisture constants are as follows:

- Saturation or water holding capacity(WHC)
- Field capacity (FC)
- Permanent wilting point (PWP)
- Available water
- Ultimate wilting point
- Moisture equivalent

1. **Saturation and maximum water holding capacity (WHC):** It is the moisture content of the soil when all the pores are filled with water. At this state, soil water potential is maximum *i.e.* zero and water held by soil is also maximum. The gravitational force may pull down a part of water and is called gravitational water or free water.
2. **Field capacity (FC):** It is the moisture content in percent of a soil on oven dry basis when it has been completely saturated and downward movement of excess water has practically ceased to negligible value. This situation usually exists within two or three days after rainfall or field irrigation. It is the upper limit of the soil moisture available to the plant growth. Soil water potential ranges from -0.1 to -0.3 bars or -01 to -03 M Pa (Mega Pascal).
3. **Permanent wilting point (PWP):** Permanent wilting point or wilting coefficient is the moisture content in percent of soil at which nearly all plants wilt and do not recover in a humid dark chamber unless water is added from an outside source. This is the lower limit of available moisture for plant growth. The moisture tension at permanent wilting point is 15 atmosphere.
4. **Available water:** The portion of water in the soil that can be readily absorbed by plant roots. It is that water which is held in the soil against a pressure of upto approximately 15 bars i.e moisture which lies between field capacity and permanent wilting point. Sandy soils has the lowest ASM while clay soil have the highest.

5. **Ultimate wilting point:** Soil moisture content at which plant die is called Ultimate wilting point.

6. **Moisture equivalent:** It is defined as the amount of water retained by a sample of initially saturated soil after being centrifuged for 1000 times that of the gravitational force for a definite period of time, usually half an hour. The moisture content when expressed as percentage moisture on oven dry basis gives the value of moisture equivalent. In medium textured soils, the value of field capacity and moisture equivalent are almost the same. In sandy and clayey soils, the values may slightly vary.

Classification of Water

A. Physical classification of soil water

Based on moisture coefficient

- **Hygroscopic water:** When water is held tightly as thin film around soil particles by adsorption forces and no longer moves in capillary pores is called hygroscopic water.
- **Capillary water:** Water retained in the soil in capillary pores (micropores) against gravity by forces of surface tension as continuous film around soil particles is called capillary water. It is available water for all biological activities.
- **Gravitational water:** Water in macropores that move downward freely under the influence of gravity beyond the root zone is called gravitational water or free water. Water that moves beyond the root zonedepth of soil is unavailable to plants.

B. Biological classification of water

Based on suitability of water for rapid growth of the plant

- **Unavailable water:** Water below wilting point. This water is not available for normal growth.
- **Available water:** Between 15-0.33 bar.
- **Superfluous water:** Free water of gravitational water.

Water Movements in Soil

Infiltration: Infiltration is the entry of fluid from one medium to another or it is the process of water entry into the soil, generally by downward flow through all or part of the soil surface is termed as infiltration.

Seepage: The lateral movement of water through soil pores or small cracks in the soil profile under unsaturated condition is known as seepage.

Permeability: It indicates the relative ease with which air and water penetrate or pass through the soil pores. Permeability of soils is generally classified as rapid, moderate and slow. Thus the permeability is rapid in coarse textured soils and slow in fine textured soils.

Deep percolation: After infilteration of water, the water moves downward into the profile. Sandy soils facilitate greater percolation as compared to clayey soils due to dominance of macro pores.

Hydraulic conductivity: The hydraulic conductivity of a soil is a measure of the soil's ability to transmit water when submitted to a hydraulic gradient. Hydraulic conductivity is defined by Darcy's Law, which, for one-dimensional vertical flow, can be written as follows:

$$V=K\,(h_1 - h_2)/L$$

Where, V is Darcy's velocity (or the average velocity of the soil fluid through a geometric cross-sectional area within the soil), h_1 and h_2 are hydraulic heads, and L is the vertical distance in the soil through which flow takes place. The coefficient of proportionality, K, in the equation is called the hydraulic conductivity.

Soil-Water Measuring Methods and Devices

A variety of methods and devices are used to measure soil-water. These methods are grouped into direct and indirect methods.

Direct methods

Gravimetric method: With the gravimetric method, soil moisture is determined by taking a soil sample from the desired soil depth, weighing it, drying it in an oven (for 48-72 hours at 105°C) and then reweighing the dry sample to determine how much water was lost. This method is simple and reliable.

Volumetric method: The volumetric water content is defined as the volume of water present in a given volume (usually/m^3) of dry soil. When it is multiplied by 100 it gives volume water percentage. This method involves collecting soil sample from the field using core sampler of known volume from representative depths in the root zone and then determining its moist and dry weights.

Spirit burning method: Soil moisture from the sample is evaporated by adding alcohol and igniting. Provided the sample is not too large, the result can be obtained in less than 10 minutes. About 1.0 ml of spirit or alcohol per g of soil sample at field capacity and 0.5 ml at permanent wilting point is adequate for evaporating the soil moisture.

Infrared moisture balance method: This method is either used for individual tests or series of tests, in practically all organic and inorganic materials directly. Infrared lamp is used for drying with moving air.

Indirect methods

Feel method: As its name implies, the feel method involves estimating soil-water by feeling the soil. This method is easy to use, and many growers schedule irrigation in this way. This method depend on the experience of the individual making the measurement.

Tensiometer

Basic principle of this method is that capillary tension increases with decrease in moisture content. A tensiometer is a sealed, airtight, water-filled tube (barrel) with a porous tip on one end and a vacuum gauge on the other, as shown in Figure. A tensiometer measures soil water suction (negative pressure), which is usually expressed as tension. This suction is equivalent to the force or energy that a plant must exert to extract water from the soil. The suction force in the porous tip is transmitted through the water column inside the tube and displayed as a tension reading on the vacuum gauge. Soil-water tension is commonly expressed in units of bars or centibars. Tensiometers work in the range from 0 to 0.8 bar.

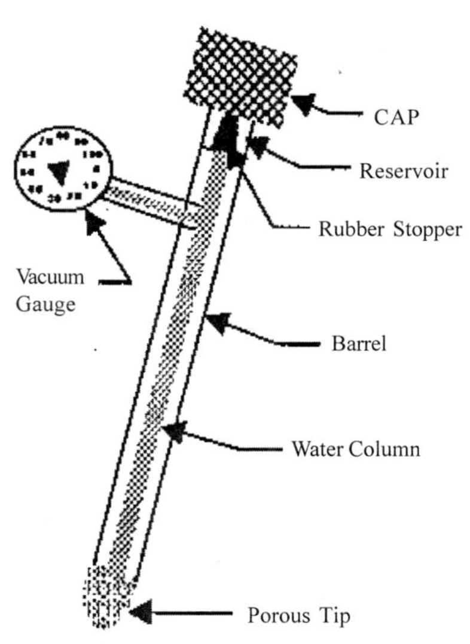

Figure: Diagram of a tensiometer.

Neutron probe method: The neutron probe uses a radioactive materials to measure soil-moisture. The basic principle of this method is that when fast moving neutrons collide with the nuclei of hydrogen atoms in the soil, they are transformed into slow neutrons. Water is the main source of hydrogen atoms in the soil, they are transformed into slow neutrons. The number of slow neutrons counts at various depths is used to estimate the amount of soil moisture content.The neutron probe consists of two main parts (i) probe unit and (ii) counting unit.

The probe contains a radioactive source for fast neutrons and a detector for slow neutrons. The source of fast neutrons is Americium Beryllium (Americium-24 and Beryllium-9). The detector has boron trifluoride. The neutron probe is inserted into the access tube by carefully lowering down the cable to the desired depth. Slow neutrons produced due to collision with hydrogen nuclei will move to the detector tube and be captured. This result in the transient breakdown of the gas in the tube. Short pulses are produced and amplified and are recorded in the counting unit.

Electrical resistance blocks: These are also called as *gypsum blocks* or *moisture blocks*. It consist of two electrodes enclosed in a block of porous material, as shown in Figure.

The block is often made of gypsum, although fiberglass or nylon is sometimes used. The electrodes are connected to insulated lead wires that extend upward to the soil surface. Resistance blocks work on the principle that water conducts electricity.When properly installed, the water suction of the porous block is in equilibrium with the soil-water suction of the surrounding soil. As the soil moisture changes, the water content of the porous block also changes. The electrical resistance between the two electrodes increases as the water content of the porous block decreases. The block's resistance can be related to the water content of the soil by a calibration curve.

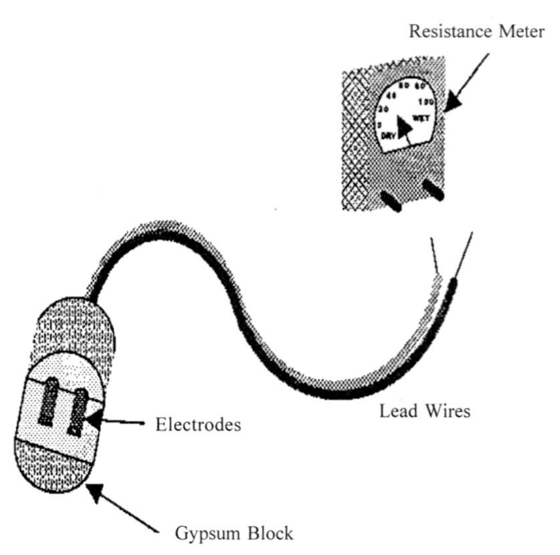

Figure : Schematic of an electrical resistance block and meter. The block is buried in the soil at one-half the effective root depth.

Time domain reflectometer (TDR): It is a new device developed to measure soil-water content. Two parallel rods or stiff wires are inserted into the soil to the depth at which the average water content is desired. The rods are connected to an instrument that sends an electromagnetic pulse (or wave) of energy along the rods. The rate at which the wave of energy is conducted into the soil and reflected back to the soil surface is directly related to the average water content of the soil. One instrument can be used for hundreds of pairs of rods. This device, just becoming commercially available, is easy to use and reliable.

Pressure plate and Pressure membrane apparatus: In laboratory, measurements of soil water potential are usually done with pressure membrane and pressure plate apparatus. It consists of ceramic pressure plates or membranes of high air entry values contained in airtight metallic chambers strong enough to withstand high pressure of 15 bars or more. The apparatus enables development of soil moisture characteristic curves over a wide range of matric potential. The porous plates are first saturated and then soil samples are placed on these plates. Soil samples are saturated with water and transferred to the metallic chambers. The chamber is closed with special wrenches to tighten the nuts and bolts with required torque for sealing it. Pressure is applied from a compressor and maintained at a desired level. It should be ensured that there is no leakage from the chamber. Water starts to flow out from saturated soil samples through outlet and continues to trickle till equilibrium against the applied pressure is achieved. Soil samples are taken out and oven dried to constant weight for determining moisture content on weight basis. Moisture content is determined against pressure values varying from 0.1 to 15 bars. The values of moisture content so obtained at a given applied pressure are used to construct soil moisture characteristic curves.

Water requirement (WR): It is the quantity of water required by a crop for its normal growth under field conditions. It is expressed in depth per unit time.

Factors affecting water requirement: There are four factors that affect the water requirement.

1. ***Climatic factors*****:** Different climatic factore like rainfall, temperature, humidity, wind velocity, photoperiod and evapo-transpiration effect the water requirement of a crop. If there is lower rainfall, high wind velocity, high temperature, low humidity, bright and longer photoperiod and higher ET, then water requirement is more.

2. ***Soil factors*****:** Different factors like organic matter, soil depth and soil colour and texture, evaporation rate all affects the water requirement of crops. Low organic matter, shallow soil depth, sandy soil, efficient drainage, dark soil colour all increase the water requirement of crops.

3. *Crop factor*

Rooting depth: Deeper root, the greater will be WR.

Plant height: Tall crop intercepts more radiation and have more evapo-transpiration and have more WR.

High leaf area: More leaf area, more is evapo-transpiration which result in more WR.

Type of culture: Rainfed crop need less water than irrigated.

4. *Management factor*: Any crop management practice aimed at improving the crop growth and high yield increases the water requirement of crop. Timely weeding and mulching minimize the evapo-transpiration resulting low WR.

Aspects of Irrigation

Since irrigation is a costly input, therefore, there is a need for achieving an efficient and economic use of irrigation water. To achieve the objective of irrigation, following aspects are:

- Time of irrigation i.e scheduling irrigation: When to irrigate.
- Depth of irrigation: How much to irrigate.
- Method of irrigation: How to irrigate or the manner in which the water is applied.

Time of irrigation i.e. scheduling irrigation: A crop should be irrigated before it receives a set back in its growth and development. The degree of soil water availability to plants tends to decline as soil water potential declines beyond a certain limit and plant may suffer in growth yield much before PWP is reached.

Frequency of irrigation depends on:

Weather: In dry weather frequent irrigation is needed.

Soil: In light soil, frequent irrigation is needed.

Water table: If there is high water table, then less frequency of irrigation is needed.

Crop: Shallow rooted crops like onion, cabbage, cauliflower *etc.* require more frequent irrigation while deep rooted crops require less frequent irrigation *eg.* Arhar, sugarcane *etc.*

Scheduling of Irrigation

Criteria for Scheduling irrigation: There are three criteria of scheduling irrigation.

i. Soil water regime approach

In this approach the available soil water held between field capacity and permanent wilting point in the effective crop root zone depth described in several ways is taken as an index or guide for determining practical irrigation schedules.

Different methods

- ***Feel and appearance of soil:*** Moisture content can be roughly estimated by taking the soil from root zone into the hand and making it into a ball. It require a lot of experience to estimate the soil moisture by this method.
- ***Depletion of the available soil moisture (DASM):*** In this method the permissible depletion level of available soil moisture in the effective crop root zone depth is commonly taken as an index or guide for scheduling irrigations to field crops. In general, for many crops scheduling irrigation's at 20–25% DASM in the soil profile was found to be optimum at moisture sensitive stages. While at other stages irrigations scheduled at 50% DASM were found optimum.
- ***Soil moisture tension:*** Soil moisture tension a physical property of film water in soil, as monitored by tensiometers at a specified depth in the crop root zone could also be used as an index for scheduling irrigations to field crops.

Optimum DASM levels for various crops

Crop	Optimum soil moisture depletion level
Maize	25–50% DASM
Sugarcane	25-65% DASM
Groundnut	25-40% DASM
Cotton	65% DASM
Sesame	50% DASM
Leafy vegetables	20% DASM
Tobacco	35% DASM
Wheat	50% DASM

ii. Climatological approach

- ***Potential evapotranspiration (PET)*:** Penmen (1948) introduced the concept of PET and he defined it as "the amount of water transpired in a unit time by short green crop of uniform height, completely covering the ground and never short of water".
- ***Lysimeter:*** Irrigation scheduling is done on the basis of soil moisture tension.
- ***Cumulative pan evaporation:*** Thus, the open pan evaporimeter being simple and as they incorporate the effects of all climatic parameters.
- ***IW:*** CPE ratio: Few examples of optimal IW/CPE ratios for important crops are given below

Crop	Optimum IW/CPE ratio
Groundnut	0.75 to 1.0
Sunflower	0.5 to 1.0
Wheat	1.0
Chick pea	0.4
Mustard	0.4
Maize	0.75 to 1.0
Sugarcane	0.5 to 1.0

iii). Plant indices approach

- ***Visual plant symptoms:*** In this method the visual signs of plants are used as an index for scheduling irrigation. For instance, plant wilting, drooping, curling and rolling of leaves in maize is used as indicators for scheduling irrigation
- ***Soil-cum-sand mini-plot technique:*** In this method one cubic metre pit is dug in the middle of the field. About 5% of sand by volume is added to the dug soil, mixed well and the pit is filled up in the normal order. Crops are grown as usual in the entire area of the field including the pit area. The plants in the pit show wilting symptoms earlier than the other plants in the remaining area. Irrigation is scheduled as soon as wilting symptoms appear on the plants in the pit.
- ***Plant population:*** In one square area, sow the seed with 4 times more seed rate than normal seed rate. Because of high plant density, plants show wilting earlier than in the rest of the crop area indicating the need scheduling of irrigation.
- Rate of growth
- Relative water content

- ***Infrared thermometry:*** It measures canopy temperature (Tc) and air temperature (Ta), when Tc – Ta is zero then there is stress.
- ***Plant indices:*** Visual signs of plant wilting can be used to schedule the irrigation like dropping, curling or rolling of leaves and change in foliage colour.
- ***Critical growth stages:*** Critical growth period is the stage or stages of growth of the crop at which moisture stress has the greatest effect on quality and quantity of yield. Therefore, time of irrigation is decided by different critical stages.

Irrigation Methods

There are four commonly used methods:

- Surface irrigation
- Sub-surface irrigation
- Sprinkler irrigation
- Drip irrigation.

Surface irrigation

It is also called gravity irrigation. Surface irrigation is the application of water by gravity flow to the surface of the field. It can be defined as the process of introducing a stream of water at the head of the field and allowing gravity and hydrostatic pressure to spread the flow over the surface throughout the field.To move forward, the flowing water must have downward slope in the direction of water flow. Field is irrigated by following methods:

1. ***Flooding:*** Water is allowed from the channel into the entire field. Less labour is required in this method.
2. ***Check basin method:*** In this method the field is divided into square or rectangular plots surrounded by bunds on all the sides and water is applied into each plot one after another. Size of check basin depends on soil texture and stream size. In this method water is applied uniformly. This method is usually practiced in nearly levelled lands and in fine textured soils with low permeability rate. The labour requirement is more in this method and land is also wasted under channels and bunds.This method is good for close grown crops. *e.g.* groundnut, wheat, pearlmillet, rice, pastures, alfalfa, clover; citrus, banana, cereals, tobacco.

3. ***Ring basin method:*** In this method a circular bund is constructed around each tree/plant or group of plants/trees to create a basin for irrigation. This method is common in orchards.

4. ***Border strip method:*** The cultivated field to be irrigated is divided into a number of long parallel strips, generally 5 to 15 m in width and 75 to 300 m in length separated by small border ridges or low dykes of about 15 cm high, laid out in the direction of the slope. Water from the channel is allowed into each strip at a time. There is no uniform water distribution in this method. This method is suitable for medium to heavy soils and not good for light soils. This method is suitable for irrigating a wide variety of close growing crops such as wheat, barley, groundnut, bajra and berseem.

5. ***Furrow irrigation:*** In this method, the flat bed surface is converted into a series of ridges and furrows running down the slope. The spacing of the furrow is decided by the spacing of row crop. The length of the furrow and slope depends on several factors *viz.,* texture, intake rate *etc*. Furrow length varies from 30-300m. Water infiltrates into the soil as it moves along the slope. The crop is usually grown on the ridges between the furrows. This method is suitable for all row crops and for crops that cannot stand in water for long periods (12-24 hours), *eg.* maize, sunflower, sugarcane, soybean, tomatoes, vegetables, potatoes, beans, citrus, grape. Uniform flat or gentle slopes are preferred for furrow irrigation. Slope along the furrows may range from 0.2 to 2%. Soils that crust easily are especially suited to furrow irrigation because the water does not flow over the ridge, and so the soil in which the plants grow remains friable. Depending upon soil conditions, there are some variations in furrow method. They are as follows:

- ***Alternate furrow system/skip irrigation:*** Water is applied to one side of each crop row.

- ***Wide spread furrow irrigation:*** Furrows of more than 2-5m apart having 2 or more than 2 crop rows are irrigated.

- ***Within row irrigation:*** Pre sowing irrigation is given and seeds are planted in the furrow itself. Subsequent irrigation is given to a shallow depth in the furrow.

- ***Short furrows:*** Short furrows of 5-6 m length and 5-6 rows are grouped into basin.

- ***Corrugation irrigation:*** It is a special type of furrow irrigation, used for broadcast crops. Corrugations are small hills pressed into the soil surface. *eg.* wheat, groundnut, setaria.

6. ***Surge irrigation***: Intermitted application of water to the field surface under gravity flow which results in 'on and of' modes of constant or variable time spans.
7. ***Cablegation***: It is an automatic method and there is water and labour saving, both. To automate the system, the plug is allowed to move downslope through the pipe at a controlled rate.

Sub-surface/Subsoil irrigation

In this method, water is applied in the form of droplet to each plant at its root zone through a network of tubing without any pressure or under low pressure. It is done through underground perforated pipes, through deep trenches at 15-30 m intervals in which water wets the root zone gradually through capillary movement. Application efficiencies vary from 30-80% depending upon conditions. Water having high salt content cannot be used.

Advantages

- Minimizes evaporation losses.
- Deep trenches can be used as drainage channels.
- Suitable where water table is shallow.
- Less weed problem.

Disadvantages

- Maintainance of pipe is difficult.
- Deep percolation through trenches.
- Interferes with the crop.

Drip/Trickle irrigation

Drip irrigation is sometimes called trickle irrigation and involves dripping water into the soil at very low rates (1-4 litres/hour) from a system of small diameter plastic pipes fitted with outlets called emitters or drippers. By capillary action, water diffuses in the soil slowly. Water is applied close to plants so that only part of the soil in which the roots grow is wetted unlike surface and sprinkler irrigation, which involves wetting the whole soil profile. This may be as, low as 30% of the volume of soil wetted by the other methods. The wetting patterns which develop from dripping water onto the soil depend on discharge and soil type. With drip irrigation water, applications are more frequent (usually every 1-3 days) than with other methods and this provides a very favourable high moisture level in the soil in which plants can flourish.

It is most suitable for row crops (vegetables, soft fruit), tree and vine crops . Generally only high value crops are considered because of the high capital costs of installing a drip system.

Drip irrigation is of two types

- Surface drip irrigation
- Sub-surface drip irrigation

Components of drip irrigation system

Pump unit: The pump unit takes water from the source and provides the right pressure for delivery into the pipe system.

Control head: The control head consists of valves to control the discharge and pressure in the entire system.

Mainlines, submains and laterals supply water from the control head into the fields.

Emitters or drippers are devices used to control the discharge of water from the lateral to the plants.They are usually spaced more than 1 metre apart with one or more emitters used for a single plant such as a tree. For row crops more closely spaced emitters may be used to wet a strip of soil.

Advantages

- Water saving upto 30-50%
- Evaporation and percolation losses are reduced
- Reduction in labour
- Reduced weed growth and weed infestation due to limited surface wetting
- Water supply is constant
- Suitable in water shortage areas
- Salt concentration is less due to high water content
- Land leveling is not required
- Fertilizer and other chemicals can be applied along with water
- Better plant growth and high yield is obtained
- Nutrient availability is increased
- Less loss of chemicals.

Disadvantages

- High initial cost of installation
- Problem in land preparation
- Clogging of emitters
- Salt accumulation is more near the plants due to less water
- Less root development.

Sprinkler Irrigation

It is a method of applying irrigation water which is similar to natural rainfall. Rate of water delivery is more than 1000 lit/hr. It operates at pressure of more than 2.5 bar and throw water as a spray upto the distance. Water is conveyed by a pump through a network of pipes, called mainlines and submains to one or more laterals and is sprayed into the air through sprinkler nozzles or perforations so that it breaksup into small water drops which fall over the land or crop surface in a uniform pattern. The main objective of a sprinkler system is to apply water as uniformly as possible to fill the root zone of the crop with water.

Sprinklers are best suited to sandy soils with high infiltration rates although they are adaptable to most soils. A good clean supply of water, free of suspended sediments, is required to avoid problems of sprinkler nozzle blockage. This method is suited for field and tree crops and water can be sprayed over or under the crop canopy. Sprinkler irrigation systems may be classified as portable, semiportable, semi permanent or permanent. They are also classified as set-move (hand-move, tow-move, side-roll and gun-type systems), solid-set or continuous move sprinkler (center-pivot, traveler and linear-move) systems.

Components of sprinkler irrigation system

Pump unit: The pump unit is usually a centrifugal pump which takes water from the source and provides adequate pressure for delivery into the pipe system.

Mainline and sometimes submainlines: are pipes which deliver water from the pump to the laterals. In some cases these pipelines are permanent and are laid on the soil surface or buried below ground. In other cases they are temporary, and can be moved from field to field. The main pipe materials used include asbestos cement, plastic or aluminium alloy.

Laterals: The laterals deliver water from the mainlines or submainlines to the sprinklers. They can be permanent but more often they are portable and made of aluminium alloy or plastic so that they can be moved easily.

Sprinklers: From the sprinklers, water falls over the soil.

Advantages

- Uniform application of water and conveyance loss is less.
- More water saving as no surface runoff is there.
- Suitable for undulating topography and steep slopes.
- Fertilizers, pesticides and herbicides can be applied with irrigation water.
- Sprinkling before frost helps in maintaining high temperature and reduces frost damage.

Disadvantages

- Initial installation cost is high.
- More labour is needed to move the pipes and sprinklers around the field.
- It is not successful in areas having high wind velocity as water is lost as evaporation.
- Causes hinderances in land preparation.
- Power requirement is higher.
- Sprinkler nozzle blockage problem is there. Therefore clean water is required.

Devices used for measuring irrigation water

Devices used for measuring irrigation water are grouped into four categories

1. Volumetric measurements
2. Velocity-area methods
    a. Float method
    b. Water meters
    c. Current meter method
    d. Tracer method

3. Measuring structures
    a. Orifices
    b. Weirs
    c. Flumes
4. Tracer methods
    a. Dilution

1. Volumetric methods (using a container)

A simple method of measuring a small irrigation stream in which the flow of water is collected in container of known volume for a measured period. Water is collected in a container (bucket or barrel) and the time required to fill is recorded with a stopwatch or with seconds on wristwatch. The rate of flow is measured as below

$$\text{Discharge rate liter/second} = \frac{\text{Volume of container (liters)}}{\text{Time required to fill (seconds)}}$$

2. Velocity area method

a) Float method

Flow rate of a stream is easily estimated by the float method. A float is any piece of material that floats on water (cork, wooden piece). A straight uniform section of the channel of about 20-25m long is selected and marked on the banks. The time required to travel the distance by the floating object is measured and the velocity is calculated. This method is simple and cheap but not accurate.

$$\text{Discharge or rate of flow} = \text{area x velocity}$$

$$Q = A \times V$$

where Q= discharge rate in m^3/sec V = velocity of flow in m/s
A = cross sectional area of channel in m^2

b) Water meters

Water meters utilize a multi blade propeller made of metal, plastic or rubber, rotating in a vertical or horizontal plane and geared to a totalizer in such a way that a numerical counter can totalize the flow in any desired volumetric units. Water meters are available for a range of sizes suiting the pipe size commonly used on the farm. There are two basic requirements for accurate operation of the water meter.

1. The pipe must flow full at all times.
2. The rate of flow must exceed the minimum for the rated range.

c) Current meter method

This method is the standard used by the U.S. Geological Survey in gauging natural streams. In this method, velocity of river or stream is measured with current meter and the discharge estimated by multiplying the mean velocity of water by the area of cross-section of the stream.

d) Tracer method

The tracer method of flow measurement requires the injection of dyes or salts into the stream at a given point. They are then detected at some point downstream, and the velocity is determined from the time required by the tracer to flow the measured distance. Because the dye or salt is often diffused throughout the stream, determine the velocity of the first and last portions of the dye and take the average. Multiply the average velocity of the dye by the cross-sectional area of the stream to obtain the flow rate. Computation is the same as for the float method.

3. Measuring structures

a) Orifices

Orifices is opening with closed perimeter and of circular or rectangular shapes through which water flows. The edges of openings are sharp and often constructed of metal. The cross sectional area of orifice is small in relation to the stream cross section. Orifice may operate under free flow or submerged flow conditions.

The types of orifices are:

It is the stream of water coming out of orifice discharge into air, the orifice is said to have free flow orifice and if the discharge is under water, it is called submerged orifice.

Submerged orifices: These are of two types:

- Those having orifices of fixed dimensions
- Those in which height of opening may be varied.

b) Weirs

A weir is a notch of a specific shape through which water may flow. Weirs are probably the most common device for measuring irrigation water in open ditches. These are used to measure the flow in an irrigation channel, or the discharge of a well or channel outlet at the source. Different types of weirs are:

1. Sharp-crested weirs: These are of three types;

- Rectangular notch or weir
- Trapezoidal or Cipolletti weir
- 90° triangular weir (V-notch).

2. Broad-crested weirs.

c) Parshall flume or (Venturi flume)

Ralph L. Parshall (1926) designed to measure water flow in Venturi flume which was gradually refined by Parshall (1950) himself as Parshall flume which is now most commonly used to measure small and medium sized streams.

Cut throat flume: Skogerboe *et al.* (1967) have developed cut throat flumes for measurement of water. Since there is no throat section (Zero throat flumes), the flumes have been given the name as out throat flumes by the designers. Discharge through the flume can occur under either free flow or submerged conditions

4.Tracer methods

It requires the injection of dyes or salts into stream at a given point. They can then be detected at some point downstream, and the velocity is determined from the time required by the tracer to flow the measured distance. As the dye or salt is often diffused through the stream, determine the velocity of the first and last portions of the dye and take the average. Multiply the velocity of the dye by the cross-sectional area of the stream to obtain the flow rate.

Drainage

Drainage means the process of removing water from the soil that is in excess of the needs of crop plants Or Drainage is the artificial removal of excess water known as free or gravitational water from the surface of land so as to create favourable soil condition for plant growth.

Causes of water logging

- Poor natural drainage of subsoil.
- High intensity of irrigation.
- Heavy seepage losses from canals, water courses etc.
- Submergence under floods and deep percolation from rainfall.
- Enclosing irrigated fields with embankments and choking natural drainage.

- Blocking of natural drainage channels by roads and railways.
- Internal stratification of irrigated soils.
- The area in low lying and excess rain cannot be carried away as a surface runoff rapidly into the drain causing water logged condition.
- There may be a hard pan that affects seepage of water to lower strata.
- There may be salts affecting water absorption by roots.

Drainage Methods

Drainage methods are divided into various categories:

Shallow/Surface drainage: Removal of free water tending to accumulate over the soil surface is called shallow drainage. It is done by digging open drains at suitable intervals and depth. Irrigation channels serves as drainage channels. These drains are used to reduce surface water and thereby reduce the pore water pressures at a much deeper level. These drains are:

- Lined or unlined ditches.
- Shallow, gravel-filled trenches.
- Herringbone or chevron shape.

Deep drainage: These perform the function of modifying the shape of the seepage flow in the slope material. These drains are:

- Deep trenches.
- Vertical bored drains - filled with sand or gravel.
- Horizontal bored drains - lined with perforated pipes.

Advantages of drainage

- Proper aeration and warmth in the root zone which are essential for proper growth.
- Increase in nutrient availability and nutrient absorption.
- Crop can get sufficient water and air.
- Improvement of soil structure.
- Beneficial bacteria that change organic matter into plant foods get necessary air and warm temperature in the soil.

- Roots go down deep and can draw up on moisture at greater depth.
- Nutrient absorption is increased.
- Seeds germinate faster and better stand of crop is obtained due to better soil conditions.
- Due to healthy growth of plants, there is increase in pest and diseases resistance.
- Weed growth can be checked by timely weeding and interculturing operations.
- Good drainage permits the removal of many toxic salts and thus, reduces damage to crops.

Disadvatages of drainage or drainage problems

Generally, Drainage problems are observed in arid areas. Causes of drainage problems are as follows:

- **Low lying area:** Water in low lying areas cannot be carried away and checks the air spaces and saturates the surface and sub soil.
- **Seepage of canals laterals or ditches:** The seepage enters underground strata at elevations higher than those of irrigated lands enter and often becomes a direct source of water logging of low lying areas.
- **Excessive use of water:** In areas where enough water is available and at cheap rates, water used is excess. This give rise to water logging condition. Wild flooding continuous irrigation or excessively long irrigation turns to promote water logging.
- **Internal stratification of irrigated soils:** The internal natural drainage of soils is often poor. The slowly permeable soils, which when irrigation water is applied, impede the percolation of the excess water. The water cannot move down fast enough and accumulate on the surface forming a thin layer and obstruct aeration.
- There may be salts affecting water absorption by roots.
- There may be a hard pan that affects seepage of water to lower strata.

Effects of Water stress on Plant growth and development

Water stress affects practically every aspects of plant growth by modifying the anatomy, morphology and physiology of the plants. Different effects of water stress in plants are mentioned below:

Germination: There is delay and poor germination of seeds as a result of which seedling emergence is slow and poor.

Protein synthesis: As synthesis of RNA and DNA is hampered, therefore protein synthesis is adversely effected. There is degradation of protein and prolene *i.e.* amino acid is accumulated in the plants.

Photosynthesis: Lack of water is the single most limiting factor in the regulation of photosynthesis. Photosynthesis is the process by which plants convert light energy, carbon dioxide and water to plant sugars. In the absence of water, no amount of light or carbon dioxide will be of use to the plant. Leaves wilt, photosynthesis mechanisms shut down and growth slows to a halt.

Wilting: Plant cells contain large vessels known as vacuoles. These structures serve a number of functions, including water storage. Well-hydrated vacuoles push against the thick cell walls making the cell rigid, or turgid. When all cells are turgid, the plant itself is firm and crisp. When a plant loses water, the vacuole contracts and cannot maintain this pressure. The cells become flaccid and the plant wilts.

Closed stomata: When plants are well-hydrated, up to 95 percent of water absorbed through the roots returns to the atmosphere by transpiration. Transpiration is the release of water vapor and oxygen gas through tiny holes in the leaves known as stomata. When moisture is limited, the stomata close to slow transpiration and conserve water. Photosynthesis cannot occur when stomata are closed, and growth stops.

Increased susceptibility to photo inhibition: Overexposure to sunlight can reduce photosynthetic efficiency. This process is known as photo inhibition.

Loss of leaves: When water is scarce, aging foliage becomes a liability to the plant.

Root damage: The root hairs primarily responsible for water uptake are also the most susceptible to damage. During the early stages of water stress, root growth increases to access water deeper in the soil. As drought continues, this extra surface area leads to loss of water. Plants respond by converting root hairs to woody cork, making them unsuitable for water transport.

Nutrient uptake : There is reduced uptake of nutrients due slow root growth and lesser solubility of nutrients under draught conditions, roots go deeper in search of moisture while nutrients remain in top soil.

Susceptibility to pest damage: Plants that are weakened from water stress are more susceptible to damage from pest species and plant disease.

Slow growth: Root damage, pest invasion and reduced photosynthesis slow down plant growth. Shoot elongation and formation, bud formation and leaf development cannot take place with inadequate resources.

Beneficial effects of Water stress

- Mild stress
- Increases rubber content
- Aromatic properties of turkish tobacco increases
- Increases in oil content in olive and soybean though the yield is decreased.

Evapo transpiration : Evapo transpiration is a combined loss of water from the soil (evaporation) and plant (transpiration) surfaces to the atmosphere through vaporization of liquid water, and is expressed in depth per unit time (for example mm/day).

Consumptive use: The term consumptive use (CU) is used to designate the sum of losses due to evaporation + transpiration from the cropped field as well as that water utilized by the plants in its metabolic activities for building up of the plant tissues. Since the water used in the actual metabolic processes is insignificant (about 1% of evapotranspiration losses) the term consumptive use is generally taken equivalent to evapotranspiration. It is expressed similar to ET as depth of water per unit time i.e., mm/day or cm/day.

Irrigation requirement (IR): It is a part of total requirement of crop exclusive of effective rainfall and soil moisture stored in the root zone or that contribute from the shallow ground water table.

IR = Water requirement (WR) – (Effective rainfall (ER) + Storage charge in soil moisture i.e. Ground water contribution (GWC)

IR = WR- (ER + GWC)

Gross irrigation requirement (GIR) : It is the total amount of water required to bring the crop root zone to field capacity (Net Irrigation Requirement) inclusive of the water required offsetting the application losses.

Net irrigation requirement (NIR): It is the amount of water that must be stored in the root zone to meet the consumptive use (CU) requirement of crop exclusive of effective rainfall.

Irrigation efficiency: It is the ratio expressed in percentage of water stored in the root zone to the water applied in the field.

IE = Water stored in the root zone /water applied x 100

Irrigation interval: It is the number of days between two successive irrigations during the period without precipitation for a given crop and field. It depends on the crop ET rate and on the available water holding capacity of the soil in the crop root zone depth. Sandy soils require in general more frequent irrigations as compared to fine textured soils.

Irrigation period: It refers to the number of days that can be allowed for applying one irrigation to that of the next in a given design area during the period of highest consumptive use of the season.

Water used efficiency: It is the ratio of economic yield to evapotranspiration

WUE = Y/ET

(Y is economic Yield)

Water requirement (WR)

It is the quantity of water required by a crop for its normal growth under field conditions. It is expressed in depth per unit time.

WR = Consumptive use (CU) + Application losses (percolation, seepage, runoff) + Water needed for special operations *i.e.* levelling, puddling, land preparation.

CU = Total of water used by a crop for its normal growth in transpiration and evaporation (99% water) *i.e.* ET and building up of tissues (1%).

Units of measuring water

Acre-inch (ac-in)

One acre-inch is the volume of water necessary to cover an acre 1 inch deep or the amount of water falling on an acre in a 1-inch rain. One acre-inch equals 3,630 cubic feet or 27,154 gallons.

Acre-foot (ac-ft)

One acre-foot is the volume of water necessary to cover an acre 1 foot deep. One acre-foot equals 43,560 cubic feet, 325,851 gallons or 12 acre-inches.

Gallons per minute (gpm)

One gallon is exactly 231 cubic inches. The gallons per minute measurement is the amount produced by a pump, stream, or pipeline in one minute.

Cubic feet per second (cfs)

One cubic foot per second is a flow of water equivalent to a stream 1 foot wide and 1 foot deep flowing at a velocity of 1 foot per second.

1 cfs = 450 gpm = 1 acre-inch per hour = 2 acre-feet per day

Head (H)

Head is a depth of water, usually in feet. It can also mean pressure; a volume of water exerts a pressure on the bottom of a container, lake or stream that is proportional to the depth of water above the bottom. One foot of water (head) exerts a pressure of 0.43 pounds per square inch on the bottom surface. Or, 1 pound per square inch is equivalent to 2.31 feet of head.

Table: Common water measurement conversions

Water Measurement Conversions	
One cubic foot per second (cfs)	40 miner's inches (Montana) 7.48 gallons per second 450 gallons per minute 646,272 gallons per day
One Montana miner's inch	0.025 cubic feet per second (cfs) 0.19 gallons per second 11.25 gallons per minute 16,158 gallons per day
1000 gallons per minute	2.2 cubic feet per second 88 miner's inches
One acre-foot	325,800 gallons 43,560 cubic feet

Terminology

pF: Schofield (1935) suggested the use of pF (by anology with the pH acidity scale), which is defined as the logarithm of the negative pressure (soil water tension or suction) head in cm of water. A tension of 10 cm of water is, thus, equal to a pF of 1. Likewise, a tension head of 1000 cm is equal to a pF of 3, and so forth.

$$pF = \log 10^{h}$$

where h= soil moisture in cm of water.

Chapter 8

Rainfed Agriculture

Out of 143 million hectares of net cropped area, about 75% is rainfed which contribute about 45% of food grains. These areas produce 55% rice, 25% wheat, 91% coarse cereals, 75% of pulses, 81% oilseeds and 69% cotton. More than 90% of sorghum, millet, groundnut and pulses is contributed from arid and semi-arid. The term rainfed farming (Dry farming) and dry land farming may be defined as conditions where rainwater is the only source of moisture for survival and growth of crops.

Dryland Agriculture may be classified as follows:

1. **Dryland farming:** It is a practice of growing profitable crops where annual rainfall is more than 750 mm but less than 1150 mm is called dryland farming.
2. **Dry farming:** It is the cultivation of crops in areas where annual rainfall is less than 750 mm and crop failure due to prolonged dry spells during crop period is very common.
3. **Rainfed farming:** It is the cultivation of crops in regions where annual rainfall is more than 1150 mm and there is very less chances of crop failures due to dry spells.

Difference between Dry farming and Dryland farming

S.No.	Characteristics	Dry farming (Rainfed farming)	Dryland farming
1.	Annual rainfall (mm)	> 750	<750
2.	Rainfall distribution	Almost certain with less inter and intra seasonal variations	Uncertain with high inter and intra seasonal variations
3.	Climate	Cold arid/ Semi arid/ Sub humid	Dry arid/ Semi arid
4.	Moisture availability to the crop	Enough	Shortage

Contd...

5.	Growing season	> 200 days	< 200 days
6.	Cropping system	Intercropping or double cropping	Single crop or intercropping
7.	Constraints	Soil erosion by wind and water	Soil erosion by water
8.	Percent contribution to arable land	42	25

Drought and types of drought

Drought is a condition of insufficient moisture supply to the plants under which they fail to develop and mature properly. It may be caused by soil, atmosphere or both or it is considered as a period of dryness affecting the earth and preventing the growth of plants (Thornthwaite 1947). He explained it as a condition when the amount of water needed for transpiration and evaporation exceeds the total amount of moisture in soil (Aridity Index).

Drought Classification

On the basis of use and duration of occurence of drought, they are classified as

1. **Meteorological drought:** The Indian Meteorological department defines as an area affected by drought if the meteorological division receives less than 75% of the total normal southwest monsoon rainfall. It is a condition where in annual precipitation is less than the normal over an area (a zone or specific location) for a prolonged period, may be month, a season or a year. Intensity of meteorological drought may be moderate (50%) or severe (>50 % deviation) depending on the deviation of the seasonal rainfall.

2. **Hydrological drought:** Prolonged meteorological drought results in hydrological drought, which leads to depletion of surface water and water level in reservoirs. If there is deficiency of water for hydrological purposes such as power generation, irrigation or for an industry which uses a large volume of water, then it is hydrological drought.

3. **Agricultural drought:** It is a situation when the rainfall and soil moisture are inadequate to meet the water requirement of crop for their healthy growth and maturity.

Drought under dry farming situation can be classified into

Early season drought	-	Late commencement of rains
Mid season drought	-	Failure of rain during crop season
Late season drought	-	Early cessation of rainy season
Permanent season drought	-	Result due to mismatching of the rainfall and water availability to the crop.

Different Institutions monitoring Dryland agriculture are

ICRISAT: International Crop Research Institute for Semi-Arid Tropics. It is established on 11 October, 1972 near Patancheru, Hyderabad.

ICARDA: International Centre for Agricultural Research in Dry Areas. It is established in 1977 at Aleppo, Syria.

Problems of Dry farming in India

The major problem which the farmers have to face very often are

- Moisture stress and uncertain rainfall
- Effective storage of rain water.

Dryland climatology

The behaviour of atmospheric phenomenon at a given place and time is defined as weather and day-to-day weather elements for a given place or region over a period is referred as climate. The systematic study of climate and its analysis and processing is referred as climatology. It plays an important role in national wealth, planning activities and decision making in human activities.

Indian agricultural zones

On the basis of occurance of annual average rainfall, there are four major zones.

i. **Arid zone:** In this zone, evaporation is greater than rainfall and annual rainfall is 30 cm or < 30 cm. Rainfall is uncertain, may or may not occur. Area covered is western Rajasthan. Drought resistant crops like Jowar, Bajra, Gram, Barley, Lentil. *etc.* are grown.

ii. **Semi-arid zone:** In this zone, evaporation is equal to rainfall and the annual rainfall is 30-75 cm. There is uncertainty of rainfall for most of the time. Area covered are Rajasthan, Haryana, Punjab, Delhi, West U.P, West M.P., Orissa, Gujarat (most of the adjoining areas of arid regions of the country). Main crops grown are Jowar, Bajra, Ragi, Barley, Wheat, Cotton, Gram, Lentil, Mustard, Arhar, Groundnut and other crops can be produced if effective use of rain water or irrigation water is done.

iii. **Sub-humid zones:** In this zone, evaporation is lesser than rainfall. The regions receive annual rainfall of about 75-150 cm. Area are Himanchal Pradesh, East Uttar Pradesh, Tarai, East Madhya Pradesh and Bihar. Crop commonly grown are Wheat, Sugarcane, Jowar, Arhar, Maize, Paddy, Mustard. More crops are grown following intensive cropping.

iv. **Humid region:** In this zone, rainfall is more than precipitation. Annual rainfall is more than 150 cm. Area covered are east and western coastal region of the country *i.e.,* Orissa, West Bengal, Assam, Andhra Pradesh. Crops like Rice, Jute, Tea are grown in this region.

Study of dryland climatology is essential in the sense that the farming in dry regions entirely depends on the climate and its components especially rainfall and its behaviour. The most widely used system of regional climatology are

1. Koeppen's classification.
2. Thornthwaite's classification.
3. Martonne's classification.
4. Maig's classification.

1. Koeppen's classification

Wlodimir Koeppen (1846 to 1940 AD) a German biologist related climate to vegetation and he gave emphasize to climatic factors like temperature, rainfall and their seasonal characteristics.

Table : Dry climatic types in Koeppen's classification

S.No.	Symbol	Climatic type
1.	BSh	Tropical steppe, semi arid & hot
2.	BSk	Mid latitude steppe, Semi arid, cool or cold
3.	BWh	Tropical desert, Arid, Hot
4.	BWk	Mid latitude desert, Arid, cool or cold

Koeppen's modified system of climate classification was based on average annual temperature (denoted as 't') and average annual values of precipitation (denoted as 'r') in ^{0}F and inches, respectively.

2. Thornthwaite classification

C.W. Thornthwaite (1899 to 1963 AD) an American climatologist classified the climate based on precipitation effectiveness (PE index) and temperature efficiency (TE).

PEI = Sum of twelve monthly values of 115 $(P/T-10)^{10/9}$

Where, P is mean monthly precipitation in inches, T is mean monthly Temperature in ^{0}F.

Temperature efficiency (TE) = Sum of twelve monthly values of (T–32/4).

Where, T is mean monthly temperature in ^{0}F.

Table: Thornthwaite classification of dry climate

S.No.	Humidity	Vegetation	PE	Temperature	TEIndex	Seasonal
1.	Semi arid (D)	Steppe	16-31	Tropical (A')	>128	Deficient in winter or deficient in all seasons.
				Mesother (B')	64-127	Deficient in winter or deficient in summer.
				Microther (C')	32-63	Deficient in all seasons.
2.	Arid (E)	Desert	<16	Tropical (A')	>128	Deficient in all seasons.
				Mesother (B')	64-127	-do-
				Microther (C')	32-63	-do-

3. Martonne's classification

Martonne (1962 AD) expressed the relationship between precipitation and temperature as Aridity Index. This term can refer to a selected period of a few days. A month, a season or whole of the year and this enables to define a dry climate more accurately.

Aridity Index I = n x p/ t + 10

Where, I = Aridity Index

n = Number of rainy days

p = Mean precipitation in mm day $^{-1}$

t = Mean temperature of the selected period in ^{0}C

On annual basis, the areas characterized by an Aridity Index less than 20 is classified as arid and between 20 and 30 is classified as semi arid.

4. Maig's classification

Maig (1953 AD) made certain refinements in Thornthwaite's system and he divided the climatic regions based on aridity as extremely arid, arid and semi arid using Thornthwaite's Moisture Index. He classified the extreme arid climate having a moisture index of -57 as outer limit, arid climate having a moisture index of -40 to -57 and semi arid climate having moisture index of -20 to -40. According to him the precipitation is not the main risk for crop production in semi arid climate and in semi arid to arid climate, it can be attempted with appropriate by farming techniques. In extreme arid climate there is no seasonal distribution of rainfall.

Conservation and efficient use of Rainwater

Soil and water are two most important natural resource available to mankind upon which depends the very survival of the living beings on this earth. Important natural resources are rainfall, soil and plants. In order to achieve good and stable yield, there are two important aspects.

i. Resource development and

ii. Their efficient use

The rainfall received in dryland areas is to be stored in the soil. Therefore, soil resources are to be improved or developed by:

- Rectifying the defects of the soil either by leveling, application of amendments etc.
- Increasing storage capacity of the soils.

Plant resources are developed by selecting or breeding drought resistant varieties suitable for arid and semi environments.

Water Conservation

It is done by following ways:

1. Water harvesting

It is also called as Runoff Concentration or Rainfall Precipitation. Water harvesting may be defined as "Collecting and storing water on the surface of the soil for subsequent use". It is a method to induce, collect, store and conserve local surface runoff for agriculture in arid and semi arid regions. In rainfed farming areas, the topography of land is undulating where the lower surface (*i.e* catchments area) acts as natural reservoir and the water is being accumulated from higher surface (*i.e.* watershed area) into it. The crop having higher and lower water requirements may be grown in catchment area and watershed area, respectively.

Types of water harvesting: These are of three types

a. ***Inter-row water harvesting***: The crop is sown in narrow strips between wide intervals that are ridged as artificial miniature watersheds. Later on, these are compacted to increase runoff to the crop rows. These are practiced in arid areas having light soils where annual rainfall does not exceed 400-450 mm.

b. ***Inter-plot or Micro-plot water harvesting***: In this type, water is harvested in the passages or furrows between the plots when rainfall is comparatively more. Runoff from the sloping area supplements rainfall for raising crop on level land.

c. ***Water harvesting in farm ponds and reservoirs***: Surface runoff from small watersheds is stored in farm ponds and reservoirs for utilization as supplemental or life saving irrigation.

2. Inducing runoff

For collection of higher amount of rainfall, runoff is induced by two methods:

Land alterations: Clearing away rocks and vegetation and compacting the soil surface can increase runoff. However, land alteration may lead to soil erosion except where slope is reduced.

Chemical treatment: Treatment of soils with chemicals that fill pores or make soil repellant to water. Same materials used for this purpose are sodium salts of silicon, latexes, asphalt and wax.

Jal shakti : A chemical which when applied (mixed) in soil improves the aeration, filteration and water holding capacity of the soil.

3. Reducing Evapotranspiration (ET) losses

About 60-75% of water is lost through evaporation. ET losses can be reduced by following ways.

A. Mulches: Mulch is any material applied on the soil surface to check evaporation and improve soil water. Mulching is helpful in soil conservation, moderation of temperature, reduction in soil salinity, weed control and improvement of soil structure.

B. Antitranspirant: Any material applied to transpiring plant surfaces for reducing water loss from the plant. These are of 4 types:

Stomatal closing type: These reduce the rate of transpiration through stomatal closing. *eg*. Phenyl Mercuric Acetate (PMA), atrazine, Abscisic acid (ABA).

Film forming type: The plastic and waxy materials, which form a thin film on the leaf surface, retard the escape of water due to formation of physical barrier. These film forming anti transpirants may be of either thin film or thick film.

Thin film forming type: Hexadeconol.

Thick film forming type: Mobileaf, Polythene S-60.

***Reflecting type*:** These are the white materials, which form a coating on the leaves and increase leaf reflectance (albedo). By reflecting the radiation, they reduce leaf temperatures and vapour pressure gradient from leaf to atmosphere and hence reduces transpiration. About 5% of Kaolin spray reduces the leaf temperature by 3-4°C and decrease in transpiration by 22 to 28 per cent. Celite and hydrated lime are also used as reflectant type of anti transpirants.

***Growth retardant type*:** These chemicals reduce shoot growth and increase root growth and thus enable the plants to reduce transpiring surface and resist drought conditions. They increase root/shoot ratio. *eg*: Cycocel – (2-chloroethyl) Trimethyl ammonium chloride (CCC), Phosphon–D, Maleic Hydrazide (MH).

There should be limited use of anti-transparents as it reduces photosynthesis

C. Wind breaks and Shelter belts : Wind breaks are any structures that obstruct wind flow and reduce wind speed. Shelter belts are planted across the direction of wind. They do not obstruct the wind flow completely. Due to reduction in wind speed, evaporation losses are reduced and more water is available for plants for example a 10-11 m tall windbreak when encountered by 45-50 km/hr wind, it reduces on windward side to 20-30 km/hr and to 10 km/hr on just Leeward side.

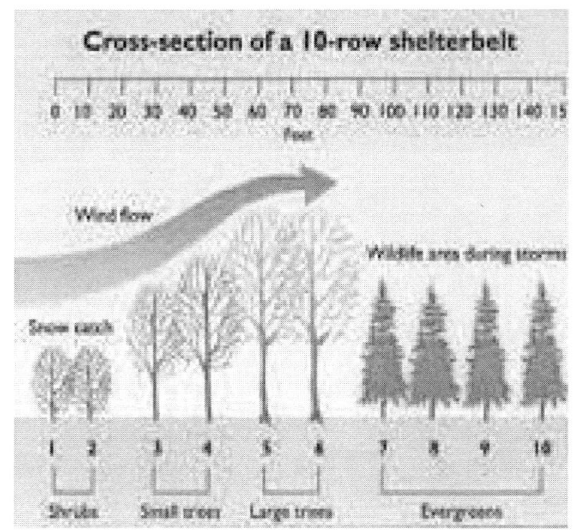

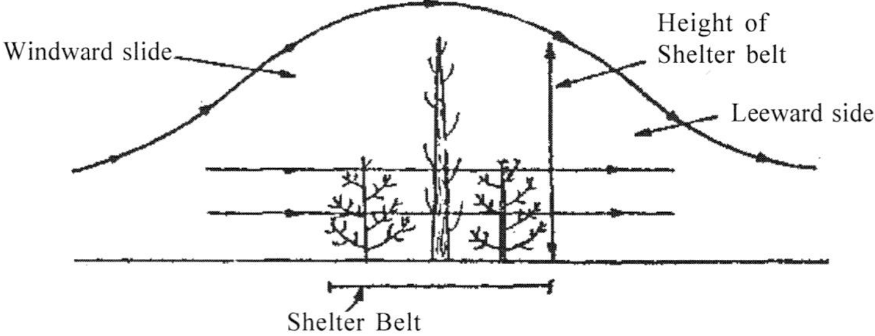

D. Weed control: Prompt weed control eliminates the competition of weeds with crops for limited soil moisture. This is the most useful measure to reduce transpiration losses.

4. Improving soil moisture storage

Addition of silt, clay, organic matter *etc.* increased field capacity also raises wilting point leading to marginal increase in available moisture holding capacity.

5. Cultivated fallowing

Soil is ploughed after taking a crop with an object to control weeds and to absorb maximum rain water. Rainwaters are preserved in the soil due to breaking of capillary tubes by ploughing.

Crop Management Practices for Rainfed areas

For Rainfed areas, different crop management practices is to be followed so that there is least damage to the crop and higher returns may be obtained.

1. **Field preparation**

   ***Deep ploughing*:** Deep ploughing is done to break hard pans, to kill weeds and to facilitate infiltration of rain water. It is done once in 3 years.

   ***Shallow tillage*:** Shallow tillage during offseason helps in weed free fields and facilitates intake of rainfall.

2. **Crop rotation:** In dry areas, only leguminous crops can be introduced which may be beneficial.
3. **Addition of organic matter:** In dryland areas, organic matter content of the soil is very low due to high temperature. By adding organic matter water holding capacity and fertility status of the soil increases.
4. **Crops and varieties**: The crops and varieties which can withstand drought are short duration, deep rooted, thick leaves and wax coated stomata.

Kharif crops: Bajra, Jowar, Minor millets (Ragi, Mandua, Kodon, Kakun, Jhangora *etc*), Soybean, Maize, Arhar, Cowpea, Groundnut, Til, Cotton, Moong, Urd, Cotton, Castor, *etc.*

Rabi crops: Barley, Wheat, Triticale, Gram, Lentil, Mustard, Linseed, Sunflower, Safflower *etc.*

Characteristics of the varieties

i. Crop must be of short duration, drought resistant and higher genetic yield potential.

ii. It should have low water requirement and deep rooted system.

iii. Leaves and stems should have waxy layer.

iv. It must be photo insensitive.

v. Leguminous crops are preferable.

6. Sowing – There are following aspects:

i. Sowing time: Sowing time should be accurate. If the crop is sown early, there is more vegetative growth and therefore conserved moisture is wasted. In late sowing, growing period is short and yield is low. When monsoon is late, then grow pearl millet, sunflower and small millets. When monsoon is in time then grow sorghum, maize.

ii. Sowing method: Row sowing, deep sowing but do not cover with too much soil.

iii. Seed treatment: Treat the seed with 0.9% Succinic acid or IAA, which can resist and tolerate moist free conditions. Succinic acid increases the rate of root growth as a result roots penetrate deeper. Seed treatment with $CaCl_2$, Boron solution, Agrosan G. N *etc.* bring drought hardiness to plants.

iv. Seed rate: Lower seed rates and higher spacing is considered best in rain-fed areas. Because of higher plant population, higher will be the water loss due to transpiration.

7. Fertilizer management: Apply bulky organic manure (FYM) @ 5-10t/ ha.

Nutrient	Wheat	Rice	Pearl millet / Sorghum
N	40-60	60	40-80
P	30	30	40

In rainfed areas crop should be fertilized with lower doses than irrigated areas.

i. **Method:** Deep placement or by the side of the row or below the seed.

- Split application of fertilizer
- Foliar spray – of urea @2-3%

8. Irrigation: Life saving irrigation at critical growth stages should be provided.

9. Cropping system: Intercropping, mixed-cropping, relay cropping, farm forestry system. Tree plantation has been found profitable in dryland agriculture.

Crop rotations are

Hyb. Napier – Hyb. Maize, G. nut

Hyb. Maize/Hyb. Jowar – Chillies

Hyb. Bajra/G. nut – Wheat (mexican)

Kh. Pulse – Wheat (mexican)

Hyb. Jowar – Gram/Wheat/Barley

G. nut – Wheat (mexican)

Kh. fodder – Wheat (mexican)

Hyb. Bajra/G. nut / Moong-Wheat (max)/Gram/Barley

Hyb. Bajra – G. nut – Tur

Hyb. Ragi – Jowar – Pulse

Cotton – Wheat (mexican)

Hy. Maize – Wheat (mex.)

Watershed Management

A watershed is defined as any spatial area from which runoff from precipitation is collected and drained through a common point or outlet. In other words, watershed is not simply the hydrological unit but also socio-political-ecological entity which plays crucial role in determining food, social and economic security and provide life support services to rural people. It is synonymous with a drainage basin or catchment area.

Watershed management implies the wise use of soil and water resources within a given geographical area so as to enable sustainable production and to minimize floods. It is the rational utilization of land and water resources for optimum production with minimum hazard to natural resources.

Classification of Watersheds

Based on the size, the watersheds may be classified as:

Micro watersheds: The size of the watershed range from few hectares to hundreds of hectares. These can be designed within the crop fields.

Small watersheds: The watershed has few thousands of hectares as drainage area.

Large watersheds: The river basins are considered as large watersheds.

Chapter 9

Questionnaire

A. Write True or False

1. Bulk density is the true density of the soil.
2. Igneous rocks are acidic in nature.
3. Adhesion is the attraction of like molecules.
4. pH is the negative logarithm of Hydrogen ion concentration.
5. Limestone is a sedimentary rocks.
6. Size of the gravel is more than 2mm.
7. Black and dark colour of soil is due to organic matter and iron.
8. Soil moisture tension at Permanent wilting point corresponds to 15 atm.
9. Micropores are the large pores present in the soil.
10. Oxidation is the removal of oxygen from the minerals.
11. Straight fertilizer contains only one nutrient.
12. Concentrated organic manure has higher nutrient content than bulky manure.
13. Bulk density of sandy soil is generally less than clay soil.
14. Acid soils are generally found in low rainfall areas.
15. Saline soils are also called as black alkali soils.
16. Vegetation increase soil erosion.
17. Jowar is erosion permitting crop.
18. Cohesion is attraction of unlike substances.

19. Size of clay is >2 mm.
20. Wheat is a kharif season crop.
21. Potato is tuber crop.
22. Citronella is medicinal crop.
23. Moong is an oilseed crop.
24. Rice is a cereal crop.
25. Cotton is a fibre crop.
26. Kharif season is from November to March.
27. Sugarbeet is an annual plant.
28. Jute belong to Tiliaceae family.
29. Potato belongs to Cruciferae family.
30. Coffee is a stimulant crop.
31. Moong is catch crop.

Answers

Write True or False

1.	False	2.	True	3.	False	4.	True	5.	True
6.	True	7.	True	8.	True	9.	False	10.	False
11.	False	12.	False	13.	False	14.	False	15.	False
16.	False	17.	True	18.	False	19.	False	20.	False
21.	True	22.	True	23.	False	24.	True	25.	True
26.	False	27.	False	28.	True	29.	False	30.	True
31.	True								

B. Fill in the blanks

1. Maize is a ....crop.
2. Cotton is a ....crop.
3. Soil contains .....% of organic matter.
4. Sedimentary rocks are ....in nature.
5. Boron is ....nutrient.
6. Barley is a ....season crop.

7. Vanadium is a ....nutrient.
8. Moisture tension at Field capacity is ....
9. Mustard belongs to ....family.
10. Annual rainfall in Dry farming is ....
11. Climate in Dryland farming is ....
12. Barley is ....season crop.
13. Cotton belongs to ....family
14. .... and ....are vegetables.
15. Berseem is a ....crop.
16. Lemon grass is a ....crop
17. Alkali soils have pH ......
18. Cohesion is attraction of ....
19. Rice belongs to .... family.
20. Maize is also called ....of cereals.

Answers

1. Cereal
2. fibre
3. 5
4. Basic
5. micro
6. Winter/Rabi
7. beneficial
8. 1/3 atm
9. Cruciferae
10. > 800 mm
11. Dry arid/semi arid
12. Rabi/winter
13. Malvaceae
14. Carrot/pumpkin.....
15. Forage
16. Medicinal
17. >8.5
18. Like particles
19. Gramineae/Poaceae
20. Queen

C. Abbreviate the following

i. ICAR
ii. IRRI
iii. FAO
iv. CIPHET
v. CAZRI
vi. IISR
vii. VPKAS
viii. DWR
ix. CIMAP
x. ICRISAT
xi. IIHR
xii. IGFRI

D. Multiple Choice Questions

1. Rainfed regions receives rainfall

   a. <750 mm b. >750 mm

   c. 2500 mm d. 1000-2500 mm

2. Acidic soil is reclaimed by

   a. Gypsum b. Lime

   c. Scraping salts d. All

3. Though not essential but beneficial effects

   a. Nitrogen b. Magnesium

   c. Sodium d. Potassium

4. Black and dark colour soil is due to

   a. Organic matter and Iron b. Selenium

   c. Silica d. Lime

5. Mineral matter in soil is

   a. 50% b. 10% c. 90% d. 45%

6. Maximum water loss from soil is due to

   a. Percolation b. Transpiration

   c. Surface runoff d. Evaporation

7. Growing grasses for 2-3 years in rotation is

   a. Monocropping b. Ley farming

   c. Relay cropping d. Terracing

8. Horizon of Eluviation (Horizon A) is also

   a. Organic matter layer b. Parent material

   c. Mineral soil layer d. Bed rock

9. Moisture tension at field capacity is

   a. 15 atmosphere b. 1/3 atmosphere

   c. 30 atmosphere d. Zero

Answers

1. b	2. b	3. c	4. a	5. d	6. c
7. b	8. c	9. a			

E. Write short notes on the following

1. Oxidation
2. Soil pH
3. Green manuring
4. Concentrated organic manures
5. Igneous rocks
6. Soil texture
7. Soil erosion
8. Mulching
9. Fertilizer
10. Bone-meal
11. Anti-transpirants
12. Wind breaks
13. Soil
14. Soil structure
15. Annuals
16. Agriculture
17. Rabi crops
18. Soil acidity
19. Weathering
20. Water harvesting
21. Green manuring
22. Vermicomposting
23. Erosion
24. Rainfed farming
25. Dryland farming
26. Bulk density
27. True density
28. Metamorphic rocks
29. Kharif crops
30. Zaid season
31. Essential elements
32. Micronutrients
33. Macronutrients
34. Acidity
35. Saline soil
36. Gypsum requirement
37. Liming
38. Medicinal plants
39. Fertilizers
40. Forage crops
41. Cohesion
42. Equivalent acidity
43. Soil profile
44. A. horizon of soil
45. Soil porosity
46. Soil texture
47. Soil consistence
48. Alluvial soils
49. Black cotton soils
50. Beneficial nutrients
51. Manures
52. Vermicomposting
53. Oil cakes
54. Blood -meal
55. Biofertilizers
56. Green manuring
57. Nitrification inhibitors
58. Fertilizer grade
59. Complex fertilizers
60. Gypsum requirement
61. Liming
62. Active acidity
63. Cation exchange capacity
64. Sodium absorption ratio
65. Water holding capacity
66. Field capacity
67. Permanent wilting point
68. Hygroscopic coefficient
69. Available water
70. Capillary water

71. Infiltration
72. Permeability
73. Electrical resistance blocks
74. pF
75. Water requirement
76. Water used efficiency
77. Irrigation efficiency
78. Meteorological drought
79. Water harvesting
80. Wind breaks
81. Antitransparents
82. Mulches
83. Shelterbelts

F. Mixed Type Questions

1. What is irrigation and the purpose of irrigation? Name the different types of irrigation.
2. Define soil and write the functions of soil.
3. What are essential nutrients? Describe "Arnon's criteria of essentiality".
4. Define soil acidity. What are different types of soil acidity?
5. What is fertilizer? Write the types of fertilizer.
6. Write methods of reclamation of soil acidity.
7. Define Erosion. Explain types of water erosion.
8. Define Irrigation. What are the basic principles of irrigation?
9. Classify crops according to botanical basis.
10. What is irrigation and the purpose of irrigation? Name the different types of irrigation.
11. In which form do plants get oxygen?
12. Molybdenum is a micronutrient. Give reason.
13. Why are carbon, oxygen, potassium and sulpher called macronutrients?
14. What is meant by 'passive absorption' of minerals by plants?
15. Name the minerals whose deficiency affects normal cell division.
16. "Deficiency of K, Ca and Mg causes necrosis of leaves". What does this statement mean ?
17. What is meant by 'passive absorption' of minerals by plants?
18. Difference between soil and sub-soil.
19. Define weathering. What are different types of weathering?
20. Explain the classification of soils in India.

21. Describe the classification of nutrients on the basis of mobility in the soil.
22. What are deficiency symptoms of phosphorus in plants?
23. Write different methods of surface irrigation.
24. What is drip irrigation. What are its advantages?
25. What are different factors affecting water requirement?
26. Define drainage. Describe different methods of drainage.
27. What are different effect of water stress in plants? Explain.
28. Define water harvesting. Explain different methods of water harvesting.
29. What are different ways of reducing evapotranspiration?

G. Differentiate between the following

a. Igneous and metamorphic rocks.
b. Saline and alkali soils.
c. Rainfed and dryland farming.
d. Bulk density and particle density.
e. Fertilizers and manures.
f. Concentrated and bulk manures.
g. Macro and micronutrients.
h. Passive and active absorption of elements.
i. Wind breaks and shelter belts.
j. Active acidity and exchange acidity.
k. Rainfed and Dryland farming.
l. Hydrological and Agricultural drought.
m. Irrigation and drainage.
n. Oxidation and reduction.
o. Rabi and Khariff season.
p. Soil profile and soil texture.
q. Gravitational water and capillary water.
r. pH and pF.
s. Water used efficiency and irrigation efficiency.

ICAR Institutions, Deemed Universities, National Research Centres, National Bureaux & Directorate/Project Directorates

Deemed Universities – 4

1. ICAR-Indian Agricultural Research Institute, New Delhi
2. ICAR-National Dairy Research Institute, Karnal
3. ICAR-Indian Veterinary Research Institute, Izatnagar
4. ICAR-Central Institute on Fisheries Education, Mumbai

Institutions – 64

1. ICAR-Central Island Agricultural Research Institute, Port Blair
2. ICAR-Central Arid Zone Research Institute, Jodhpur
3. ICAR-Central Avian Research Institute, Izatnagar
4. ICAR-Central Inland Fisheries Research Institute, Barrackpore
5. ICAR-Central Institute Brackishwater Aquaculture, Chennai
6. ICAR-Central Institute for Research on Buffaloes, Hissar
7. ICAR-Central Institute for Research on Goats, Makhdoom
8. ICAR-Central Institute of Agricultural Engineering, Bhopal
9. ICAR-Central Institute for Arid Horticulture, Bikaner
10. ICAR-Central Institute of Cotton Research, Nagpur
11. ICAR-Central Institute of Fisheries Technology, Cochin
12. ICAR-Central Institute of Freshwater Aquaculture, Bhubneshwar
13. ICAR-Central Institute of Research on Cotton Technology, Mumbai
14. ICAR-Central Institute of Sub Tropical Horticulture, Lucknow
15. ICAR-Central Institute of Temperate Horticulture, Srinagar
16. ICAR-Central Institute on Post harvest Engineering and Technology, Ludhiana
17. ICAR-Central Marine Fisheries Research Institute, Kochi
18. ICAR-Central Plantation Crops Research Institute, Kasargod
19. ICAR-Central Potato Research Institute, Shimla
20. ICAR-Central Research Institute for Jute and Allied Fibres, Barrackpore
21. ICAR-Central Research Institute of Dryland Agriculture, Hyderabad
22. ICAR-National Rice Research Institute, Cuttack
23. ICAR-Central Sheep and Wool Research Institute, Avikanagar, Rajasthan
24. ICAR- Indian Institute of Soil and Water Conservation, Dehradun
25. ICAR-Central Soil Salinity Research Institute, Karnal
26. ICAR-Central Tobacco Research Institute, Rajahmundry
27. ICAR-Central Tuber Crops Research Institute, Trivandrum
28. ICAR-ICAR Research Complex for Eastern Region, Patna
29. ICAR-ICAR Research Complex for NEH Region, Barapani
30. ICAR-Central Coastal Agricultural Research Institute, Ela, Old Goa, Goa

31. ICAR-Indian Agricultural Statistics Research Institute, New Delhi
32. ICAR-Indian Grassland and Fodder Research Institute, Jhansi
33. ICAR-Indian Institute of Agricultural Biotechnology, Ranchi
34. ICAR-Indian Institute of Horticultural Research, Bengaluru
35. ICAR-Indian Institute of Natural Resins and Gums, Ranchi
36. ICAR-Indian Institute of Pulses Research, Kanpur
37. ICAR-Indian Institute of Soil Sciences, Bhopal
38. ICAR-Indian Institute of Spices Research, Calicut
39. ICAR-Indian Institute of Sugarcane Research, Lucknow
40. ICAR-Indian Institute of Vegetable Research, Varanasi
41. ICAR-National Academy of Agricultural Research & Management, Hyderabad
42. ICAR-National Institute of Biotic Stresses Management, Raipur
43. ICAR-National Institute of Abiotic Stress Management, Malegaon, Maharashtra
44. ICAR-National Institute of Animal Nutrition and Physiology, Bengaluru
45. ICAR-National Institute of Research on Jute & Allied Fibre Technology, Kolkata
46. ICAR-National Institute of Veterinary Epidemiology and Disease Informatics, Hebbal Bengaluru
47. ICAR-Sugarcane Breeding Institute, Coimbatore
48. ICAR-Vivekananda Parvatiya Krishi Anusandhan Sansthan, Almora
49. ICAR-Central Institute for Research on Cattle, Meerut, Uttar Pradesh
50. ICAR-National Institute of High Security Animal Diseases, Bhopal
51. ICAR-Indian Institute of Maize Research, New Delhi
52. ICAR- Central Agroforestry Research Institute, Jhansi
53. ICAR-National Institute of Agricultural Economics and Policy Research, New Delhi
54. ICAR-Indian Institute of Wheat and Barley Research, Karnal
55. ICAR-Indian Institute of Farming Systems Research, Modipuram
56. ICAR-Indian Institute of Millets Research, Hyderabad
57. ICAR-Indian Institute of Oilseeds Research, Hyderabad
58. ICAR-Indian Institute of Oil Palm Research, Pedavegi, West Godawari
59. ICAR-Indian Institute of Water Management, Bhubaneshwar
60. ICAR-Indian Institute of Rice Research, Hyderabad
61. ICAR-Central Institute for Women in Agriculture, Bhubaneshwar
62. ICAR-Central Citrus Research Institute, Nagpur
63. ICAR-Indian Institute of Seed Research, Mau
64. ICAR-Indian Agricultural Research Institute, Hazaribag, Jharkhand

National Research Centres – 15

1. ICAR-National Research Centre for Banana, Trichi
2. ICAR-National Research Centre for Grapes, Pune
3. ICAR-National Research Centre for Litchi, Muzaffarpur
4. ICAR-National Research Centre for Pomegranate, Solapur

5. ICAR-National Research Centre on Camel, Bikaner
6. ICAR-National Research Centre on Equines, Hisar
7. ICAR-National Research Centre on Meat, Hyderabad
8. ICAR-National Research Centre on Mithun, Medziphema, Nagaland
9. ICAR-National Research Centre on Orchids, Pakyong, Sikkim
10. ICAR-National Research Centre on Pig, Guwahati
11. ICAR-National Research Centre on Plant Biotechnology, New Delhi
12. ICAR-National Research Centre on Seed Spices, Ajmer
13. ICAR-National Research Centre on Yak, West Kemang
14. ICAR-National Centre for Integrated Pest Management, New Delhi
15. National Research Centre on Integrated Farming (ICAR-NRCIF), Motihari

National Bureaux – 6

1. ICAR-National Bureau of Plant Genetics Resources, New Delhi
2. ICAR-National Bureau of Agriculturally Important Micro-organisms, Mau, Uttar Pradesh
3. ICAR-National Bureau of Agricultural Insect Resources, Bengaluru
4. ICAR-National Bureau of Soil Survey and Land Use Planning, Nagpur
5. ICAR-National Bureau of Animal Genetic Resources, Karnal
6. ICAR-National Bureau of Fish Genetic Resources, Lucknow

Directorates/Project Directorates – 13

1. ICAR-Directorate of Groundnut Research, Junagarh
2. ICAR-Directorate of Soybean Research, Indore
3. ICAR-Directorate of Rapeseed & Mustard Research, Bharatpur
4. ICAR-Directorate of Mushroom Research, Solan
5. ICAR-Directorate on Onion and Garlic Research, Pune
6. ICAR-Directorate of Cashew Research, Puttur
7. ICAR-Directorate of Medicinal and Aromatic Plants Research, Anand
8. ICAR-Directorate of Floricultural Research, Pune, Maharashtra
9. ICAR-Directorate of Weed Research, Jabalpur
10. ICAR-Project Directorate on Foot & Mouth Disease, Mukteshwar
11. ICAR-Directorate of Poultry Research, Hyderabad
12. ICAR-Directorate of Knowledge Management in Agriculture (DKMA), New Delhi
13. ICAR-Directorate of Cold Water Fisheries Research, Bhimtal, Nainital

The Agricultural Universities 63 are major partners in growth & development of Agricultural Research and Education under National Agricultural Research System. For any relevant information, you may contact any of the State Agricultural Universities

S.No.	Name, email & website	Address	Telephone/Fax No.
1	Acharya NG Ranga Agricultural University Website: http://www.angrau.net Email: angrau_vc@yahoo.com raghuvardhanreddy_s@rediffmail.com	Adminstrative Office, Rajendra Nagar, Hyderabad-500030 Andhra Pradesh	040-24015035, 24013095 Fax: 040-24015031
2	Agriculture University Jodhpur Website: http://www.au-ju.org Email: vcunivag@gmail.com	Mandor, Jodhpur-342304	0291--2570711 Fax: 0291-2570710
3	Agriculture University Kota Website: http://aukota.org Email: vcaukota@gmail.com	Borkhera, Kota-324001	0744--2207212
4	Anand Agricultural University Website: http://www.aau.in Email: vc@aau.in, vc_aau@yahoo.com	Anand-388110, Gujarat	02692-261273 Fax: 02692-261520
5	Assam Agricultural University Website: http://www.aau.ac.in Email: vc@aau.ac.in, kmbujarbaruah@rediffmail.com	Jorhat-785013, Assam	0376-2340013 Fax: 0376-2340001
6	Bidhan Chandra Krishi Viswavidyalaya Website: http://www.bckv.edu.in Email:bckvvc@gmail.com sarojsanyal@yahoo.co.in	Mohanpur, Nadia-741252, West Bengal	033-25879772, 03473-222666 Fax:03473-222275
7	Bihar Agricultural University Website: http://www.bausabour.ac.in Email: vcbausabour@gmail.com	Sabour, Bhagalpur-813210, Bihar	0641-2452606 Fax:0641-2452604

8	Birsa Agricultural University Website: http://www.baujharkhand.org Email: vc_bau@rediffmail.com	Kanke, Ranchi-834006, Jharkhand	0651-2450500 Fax:0651-2450850
9	Central Agricultural University Website:http://www.cau.org.in Email: snpuri04@yahoo.co.in, snpuri@rediffmail.com	P.O. Box 23, Imphal-795004, Manipur	0385-2415933 Fax:0385-2410414
10	Chandra Shekar Azad University of Agriculture & Technology Website:http://www.csauk.ac.in Email: vc@csauk.ac.in	Kanpur-208002, Uttar Pradesh	0512-2534155 Fax:0512-2533808
11	Chaudhary Charan Singh Haryana Agricultural University Website:http://www.hau.ernet.in Email: vc@hau.ernet.in	Hisar-125004, Haryana	01662-231640, 284301 Fax:01662-234952
12	CSK Himachal Pradesh Krishi Vishvavidyalaya Website: http://www.hillagric.ac.in Email: vc@hillagric.ac.in	Palampur-176062, Himachal Pradesh	01894-230521 Fax:01894-230465
13	Chhattisgarh Kamdhenu Vishwavidyalaya Website: http://cgkv.ac.in Email:	Anjora, Durg, Chhattisgarh	
14	Dr Balasaheb Sawant Konkan Krishi Vidyapeeth Website: www.dbskkv.org Email: vcbskkv@yahoo.co.in	Dapoli Distt, Ratnagiri-415 712 Maharashtra	02358-282064 Fax:02358-282074
15	Dr Panjabrao Deshmukh Krishi Vidyapeeth Website: http://www.pdkv.ac.in Email: vc@pdkv.ac.in	Krishinagar,Akola-444104 Maharashtra	0724-2258365 Fax: 0724-2258219

16	Dr Yashwant Singh Parmar Univ of Horticulture & Forestry Website: http://www.yspuniversity.ac.in Email: vc@yspuniversity.ac.in, vcuhf@yahoo.com	Solan, Nauni-173230 Himachal Pradesh	01792-252363 Fax:01792-252242
17	Dr YSR Horticultural University Website: http://www.drysrhu.edu.in Email: vc@drysrhu.edu.in	Adminstrative office, Venkataramannagudem, PB No. 7, West Godavari Dist. Tadepalligudem-534101, Andhra Pradesh	08818-284311 Fax: 08818-284223 05944-233333,233500
18	Govind Ballabh Pant University of Agriculture & Technology Website: http://www.gbpuat.ac.in Email:vcgbpuat@gmail.com	Pantnagar-263145, Distt Udham Singh Nagar, Uttaranchal	Fax: 05944-233500
19	Guru Angad Dev Veterinary and Animal Science University Website: http://www.gadvasu.in Email:vijay_taneja@hotmail.com vcgadvasu@gmail.com	Ludhiana - 141004, Punjab	0161-2553320,2553360 Fax:0161-2553340
20	Indira Gandhi Krishi Vishwavidyalaya Website: www.igau.edu.in Email: vcigkv@gmail.com	Krishak Nagar, Raipur-492006 Chhattisgarh	0771-2443419 Fax: 0771-2442302, 2443121
21	Jawaharlal Nehru Krishi Viswavidyalaya Website: http://www.jnkvv.nic.in Email: vst.vcjnkvv@gmail.com	Krishi Nagar, Adhartal Jabalpur-482004, Madhya Pradesh	Fax:0761-2681389 0761-2681706
22	Junagadh Agricultural University Website: http://www.jau.in Email: vc@jau.in	Univ. Bhavan, Motibagh Junagadh-362001, Gujarat	0285-2671784 Fax:0285-2672004

23	Karnataka Veterinary Animal and Fisheries Sciences University Website: http://www.kvafsu.kar.nic.in Email: vckvafsu@yahoo.co.in dekvafsub@yahoo.com, sskvafsu@yahoo.co.in	Nandinagar, PB No 6, Bidar-585401 Karnataka	08482-245264 Fax:08482- 245107
24	Kerala Agricultural University Website: http://www.kau.edu Email: vc@kau.in, vicechancellorkau@gmail.com	Vellanikara, Trichur-680656, Kerala	0487-2371928, 2370034, 2438001 Fax:0487-2370019
25	Kerala University of Fisheries & Ocean Studies Website: http://www.kufos.ac.in Email: kurup424@gmail.com	Papangad, Kochi-682506 Kerala	0487-2370117, 2703781, 2700964
26	Kerala Veterinary and Animal Sciences University Website: http://www.kcasu.ac.in Email: vc@kvasu.ac.in, vc.vetuny@gmail.com	Liason Office, Directorate of Dairy Development, Pattom Thiruvananthapuram-695004, Kerala	0471-2550058 Fax: 0471-2550480
27	Lala Lajpat Rai University of Veterinary & Animal Sciences Website: http://www.llruvas.edu.in/ Email: vc@llruvas.edu.in	Hisar, Haryana	01662 289332 01662- 272002
28	Nanaji Deshmukh Veterinary Science University Website: http://www.mppcvv.org Email: vcnduvs@gmail.com	South Civil Lines, Jabalpur-482001, Madhya Pradesh	0761-2678007 Fax: 0761-2620783
29	Maharana Pratap Univ. of Agriculture & Technology Website: http://www.mpuat.ac.in Email: vc@mpuat.ac.in	Udaipur, Rajasthan-313001	0294-2471101 Fax:0294-2470682
30	Maharashtra Animal Science & Fishery University Website: http://www.mafsu.in Email: vcmatsu@gmail.com cadaba_prasad@yahoo.co.in	Seminary Hills, Nagpur-440006 Maharashtra	0712-2511088 Fax:0712-2511282

31	Mahatma Phule Krishi Vidyapeeth Website: http://mpkv.mah.nic.in Email: vcmpkv@rediffmail.com	Rahuri-413722, Maharashtra	02426-243208 Fax:02426-243302
32	Manyavar Shri Kanshiram Ji University of Agriculture and Technology Website: http://www.mskjuat.edu.in/ Email: vc.mskjuat@gmail.com	Banda - 210001, Uttar Pradesh	05192-221605 Fax:02426-243302
33	Narendra Deva University of Agriculture & Technology Website:http://www.nduat.ernet.in Email:vc_nduat2010@yaho.co.in	Kumarganj, Faizabad -224229 Uttar Pradesh	05270-262097, 262161 Fax:05270-262097
34	Navsari Agricultural University Website: http://www.nau.in Email: vc_2004@yahoo.co.in	Navsari-396450 Gujarat	02673-283869 Fax:02673-284254
35	Orissa Univ. of Agriculture & Technology Website: http://www.ouat.ac.in Email: ouat_dproy@yahoo.co.in, bsenapati1942@yahoo.com	Bhubaneshwar-751003, Orissa	0674-2397700 Fax:0674-2397780
36	Prof. Jayashankar Telangana State Agricultural University Website:www.pjtsau.ac.in Email: vcpjtsau@gmail.com	Admn. Office: Rajendranagar Hyderabad-500 030	40-24015122 Fax:040-24018653 (M):09849029245
37	Punjab Agricultural University Website: http://www.pau.edu Email: vcpau@pau.edu	Ludhiana-141004, Punjab	0161-2401794 Fax: 0161-2402483
38	Rajasthan University of Veterinary and Animal Sciences Website: http://rajuvas.org Email: vcrajuvas@gmail.com	Bijey Bhavan Place Complex (Pt Deen Dayal circle) Bikaner-334006 Rajasthan	0151-2543419 Fax: 0151-2549348

39	Rajendra Agricultural University Website: http://www.pusavarsity.org.in Email: : vcrau@sify.com	Pusa, Samastipur-848125 Bihar	06274-240226 Fax:06274-240255
40	Rajmata Vijayraje Sciendia Krishi Vishwa Vidyalaya Website: http://www.rvskvv.nic.in Email: vcrvskvv@gmail.com	Race Cource Road, Gwalior-474002 Madhya Pradesh	0751-2467673 Fax: 0751-2467673
41	Rani Laxmi Bai Central Agricultural University Website: http://www.rlbcau.ac.in Email: : ddgedn@icar.org.in	Jhansi, Uttar Pradesh	011-25841760 Fax:011-25843932
42	Sardar Vallabhbhai Patel University of Agriculture and Technology Website: http://www.svbpmeerut.ac.in Email:vc_agunivmeerut@yahoo.com	Modipuram, Meerut-250110 Uttar Pradesh	0121-2888522, Fax:0121-2888505
43	Sardarkrushinagar-Dantiwada Agricultural University Website: http://www.sdau.edu.in Email: vc@sdau.edu.in	Sardar Krushinagar, Distt Banaskantha, Gujarat-385506	02748-278222 Fax:02748-278261
44	Sher-E-Kashmir Univ of Agricultural Sciences & Technology Website: http://www.skuast.org Email: vc@skuast.org	Railway Road, Jammu 18009, J&K	0191-2263714 Fax:0191-2262073
45	Sher-E-Kashmir Univ of Agricultural Sciences & Technology of Kashmir Website: http://www.skuastkashmir.ac.in Email: vc@skuastkashmir.ac.in, skuastkvc@gmail.com	Shalimar Campus, Shrinagar-191121, Jammu & Kashmir	0194-2464028, 2462159 Fax:0194-2462160, 2461543

46	Sri Karan Narendra Agriculture University Website:http://sknau.ac.in Email:nsrdsr@gmail.com, vc@sknau.ac.in	Jobner-303329	0877-2248986 Fax:0877-2249222
47	Sri Venkateswara Veterinary University Website:http://www.svvu.edu.in Email:prabhakarvrao@yahoo.com	Admn office, Regional Library Building Tirupati-517502	0877-2248986 Fax:0877-2249222
48	Swami Keshwanand Rajasthan Agricultural University Website: http://www.raubikaner.org Email: vcrau@raubikaner.org	Bikaner-334006, Rajasthan	0151-2250443, 2250488 Fax:0151-2250336
49	Tamil Nadu Agricultural University Website: http://www.tnau.ac.in Email: vc@tnau.ac.in	Coimbatore-641003, Tamil Nadu	0422-2431788, 2431672 Fax:0422-2431672
50	Tamil Nadu Fisheries University Website: http://www.tnfu.org.in Email:	Nagapattinam-611 003, Tamil Nadu	04365-253011
51	Tamil Nadu Veterinary & Animal Science University Website: http://www.tanuvas.ac.in Email:vc@tanuvas.org.in, karanmgk@gmail.com	Chennai-600051, Tamilnadu	044-25551574, 25551575 Fax:0444-225551576
52	University of Agricultural Sciences, Bangalore Website: http://www.uasbangalore.edu.in Email:vcuasb1964@gmail.com, vc@uasbangalore.edu.in	GKVK, Bengaluru-560065, Karnataka	080-23332442 Fax:080-23330277
53	University of Agricultural Sciences, Dharwad Website: http://www.uasd.edu Email: vc_uasd@rediffmail.com	Dharwad-580005, Karnataka	0836-2447783, Fax:0836-2448349

54	University of Agricultural Sciences, Shimoga Website: http://www.uasbangalore.edu.in/asp/agriShimoga.asp Email:	Shimoga, Karnataka	
55	University of Horticultural Sciences Website: http://uhsbagalkot.edu.in/ Email: dandinbnm@gmail.com	Sector No 60 Navanagar Bagalkot-587102 Karnataka	08354 201310 Fax: 08354 235152
56	University of Agricultural Sciences Website: http://www.uasraichur.edu.in Email: vcuasraichur10@rediffmail.com	PB 329, Raichur-584101 Karnataka	08532-221444 Fax: 08532- 220444
57	UP Pandit Deen Dayal Upadhaya Pashu Chikitsa Vigyan Vishwa Vidhyalaya evam Go Anusandhan Sansthan Website: http://www.upvetuniv.edu.in Email: singhambika1945@gmail.com, duvasuvc@gmail.com	Mathura-281001 , Uttar Pradesh	0565-2470199 Fax:0565-2404819
58	Uttarakhand University of Horticulture and Forestry Website: http://www.uuhf.ac.in/ Email:	Pauri Garhwal, Uttarakhand	09412051976
59	Uttar Banga Krishi Viswavidyalaya Website:http://www.ubkv.ac.in Email: vcubkv@gmail.com, vcubkv@rediffmail.com	P.O. Pundibari, Dist. Coach Bihar-736165,West Bengal	03582-270141, 270013 Fax:03582-270249
60	Vasantrao Naik Marathwada Agricultural University Website: http://www.mkv2.mah.nic.in Email: vcmau@rediffmail.com	Parbhani-431402, Maharashtra	02452-223002 Fax:02452-223582

61	West Bengal University of Animal & Fishery Sciences Website:http://www.wbuafscl.ac.in Email: wbuafs@wb.nic.in	68 KB Sarani, Kolkata-700037 West Bengal	033-25563450 Fax:033-25571986
62	Sri Konda Laxman Telangana State Horticultural University Email: vcskltshu@gmail.com registrarskltshu@gmail.com	Rajendra Nagar Campus, Hyderabad	Vice-Chancellor 040-24014301 Registrar 09908216634 (M)
63	Kamdhenu University Email: vc.kamdhenu.university@gmail.com	Karmayogi Bhavan, Block-1, B1-wing 4th Floor, Room No.414, Sector-10-A Gandhinagar-382010	Vice-Chancellor 8980009803 (M) Registrar 9725001060 (M)

Index

A

B

C